ALSO BY GURU MADHAVAN

Applied Minds: How Engineers Think

Wicked Problems: How to Engineer a Better World

Link Aviation Division   Mar 18 - 1935

Praise for Guru Madhavan's *Wicked Problems*

"The heroes of Guru Madhavan's compact book about the logical habits of engineers are not the usual suspects of the iPhone era . . . the author takes us back to an earlier time so that we can witness the solving of problems that have long since gone away."
—Jon Gertner, *The Wall Street Journal*

"Madhavan champions the importance of looking at seemingly technical problems through a wider lens—such as the business, policy and social aspects." —Susan Krumdieck, *Nature*

"Guru Madhavan's extraordinarily creative ideas and imagination best categorize him as an engineer's engineer."
—Rita Colwell, former director of the National
Science Foundation, winner of the National Medal
of Science, and author of *A Lab of One's Own*

"Profound, readable, and above all humane—a substantively and stylishly engineered book."
—Tim Harford, author of *The Data Detective*

"A wonderful and energizing book. . . . It is an espresso machine for the imagination." —Rory Sutherland, author of *Alchemy*

"In this wise, wide-ranging, and wonderful work, Guru Madhavan catalogs the kinds of engineering that have shaped the modern world—often not in the ways expected. And all too often such engineering is carried out by those who persist and persevere in the background to maintain our world from collapsing, and who are ignored without credit or reward, a point seldom appreciated by those who allocate funding and celebrate excellence."
—Don Norman, former vice president at Apple and author of
The Design of Everyday Things and *Design for a Better World*

The Blue Box

Ed Link and the Birth of Modern Aviation

Excerpted from *Wicked Problems*

Guru Madhavan

W. W. NORTON & COMPANY
Independent Publishers Since 1923

Copyright © 2024 by Guru Madhavan

Excerpted from *Wicked Problems: How to Engineer a Better World* by Guru Madhavan, published by W. W. Norton & Company, Inc., in 2024.

Manufacturing by Versa Press
Book design by Lovedog Studio
Production manager: Gabi Montgomery

Library of Congress Cataloging-in-Publication Data is available.

ISBN 978-1-324-11794-0

W. W. Norton & Company, Inc., 500 Fifth Avenue, New York, NY 10110
www.wwnorton.com

W. W. Norton & Company Ltd., 15 Carlisle Street, London W1D 3BS

Authorized EU representative: EAS, Mustamäe tee 50, 10621 Tallinn, Estonia

1 2 3 4 5 6 7 8 9 0

To Binghamton

Sept. 29, 1931.   1,825,462

E. A. LINK, JR
COMBINATION TRAINING DEVICE FOR STUDENT
AVIATORS AND ENTERTAINMENT APPARATUS
Filed March 12, 1930   4 Sheets—Sheet 1

Fig.1.

Fig.3.

Fig.2.

INVENTOR
EDWIN A. LINK, JR.
BY
ATTORNEY

Contents

Prologue: **Airplane Mode**   *1*

Chapter 1:   **Pitch**   *11*

Chapter 2:   **Roll**   *31*

Chapter 3:   **Yaw**   *55*

Chapter 4:   **Surge**   *75*

Chapter 5:   **Sway**   *95*

Chapter 6:   **Heave**   *115*

Epilogue: **Thinking Inside the Box**   *135*

Acknowledgments   *145*

Sources and Resources   *147*

Index   *165*

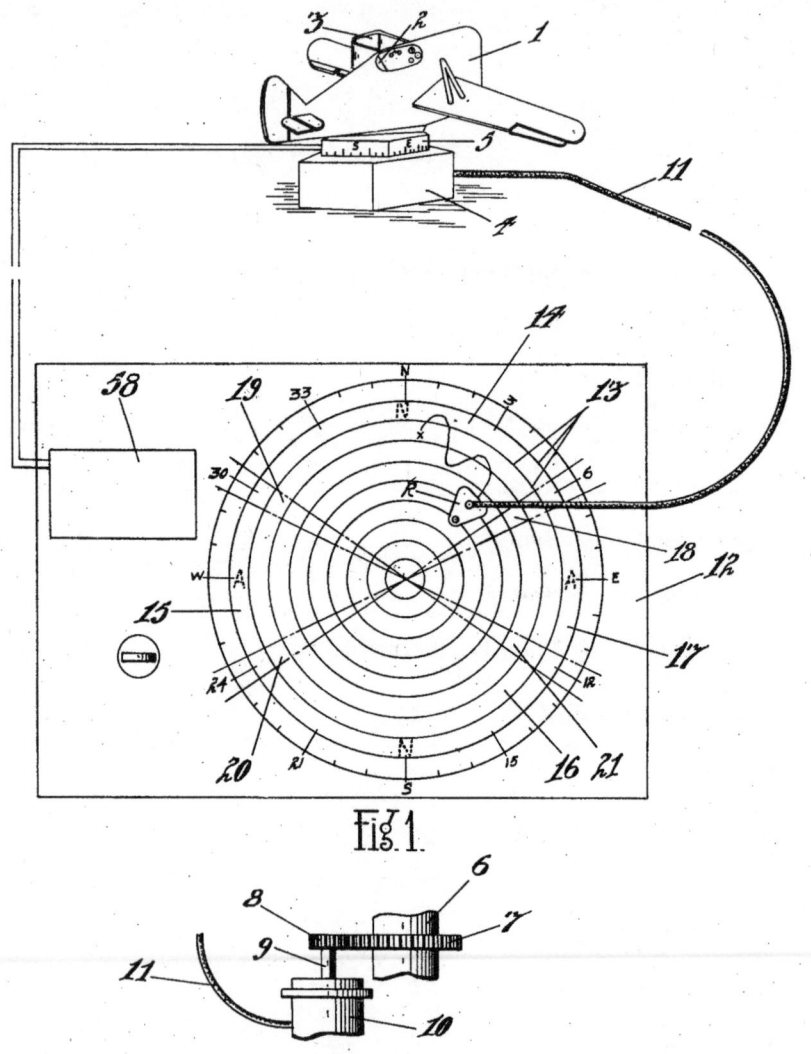

Fig.1

Fig.2

INVENTOR.
EDWIN A. LINK, JR.

BY Philip S. Hopkins

ATTORNEY.

Airplane Mode

Say, it's only a paper moon
Sailing over a cardboard sea
But it wouldn't be make-believe
If you believed in me

—Harold Arlen, Yip Harburg, and Billy Rose

IN LATE MAY 1941, 50 BRITISH CADETS SET OFF FOR THE United States. Each had received a small blue book of instructions. "You are going to America as guests," the book opened, describing the "great, friendly, yet different nation." The trainees were warned not to speak out about what was wrong with their hosts and their country.

The group boarded *Pasteur*, a lice-infested liner once owned by the French. The Atlantic was rough. The seasick cadets slept on hammocks and bedraggled bunks, fearing German torpedo attacks. They arrived filthy and famished at the dispatching station in Halifax, Nova Scotia, and were deloused and fed. As they embarked on a long train ride, the cadets were issued thick long clothes, overcoats, and woolen underwear and socks. One problem: it was June, and they were going to Texas. Their destination was the No. 1 British Flying Training School.

The town of Terrell housed one of six schools created under the American Lend-Lease Act dedicated to Royal Air Force pilot training. The others were in Lancaster, California; Miami, Oklahoma; Mesa, Arizona; Clewiston, Florida; and Ponca City, Oklahoma, along with a short-lived summer school in Sweetwater, Texas. Each site contained an airfield, a ground school, a hangar, and a maintenance facility. The Royal Air Force set the course syllabus to train 500 novice students for flying badges at each site each year.

Six months earlier, President Franklin Roosevelt had sought public support for the lend-lease proposal. He viewed the American obligation to Great Britain as a good neighbor lending a hose to stop the fire of World War II. "Manufacturers of watches, farm implements, linotypes, cash registers, automobiles, sewing machines, lawn mowers and locomotives are now making fuses, bomb packing crates, telescope mounts, shells, pistols and tanks. But all our present efforts are not enough," Roosevelt said in his December 1940 fireside chat. "We must have more ships, more guns, more planes—more of everything." The United States, he stressed, must be "the great arsenal of democracy." Congress passed the bill in March 1941.

The lend-lease project eventually cost nearly $50 billion (approaching a trillion in today's dollars), with Great Britain receiving the lion's share of matériel. The US War Department authorized the British Air Ministry to operate the contract flying schools. The British Flying Training Schools rekindled a special military relationship between the countries, as captured in the inscription of Terrell's coat of arms: "The seas divide us, the skies unite us."

British cadets were captivated by open country and abundance: from breakfast eggs to Buicks and from Coca-Cola to city lights. They relished corn on the cob and grits, what they knew previously as animal feed and as a sand-gravel mix. They learned "squash" meant vegetable, not a game, and "gasoline" was petrol, as words ping-ponged between British accents and southern drawl. The locals gave cadets replacement cotton clothing to avoid heat rash from woolen garments, and sunglasses to fly in blinding brightness. As historian Tom Killebrew describes in *The Royal Air Force in*

Texas, the cadets also witnessed the effects of segregation. Some had never seen Black people and didn't know whether to adhere to or rebel against their hosts' views. The cadets paid the Black custodial staff 50 cents weekly to make up their beds and a nickel to spit-shine shoes.

The Americans charged the British $21.60 per hour of primary training, which included ground classes on the basics of airmanship, armaments, and aircraft maintenance. Students confronted a mosaic of sensations as they learned takeoffs, turns, and touchdowns en route to their first solo flight. Advanced training offered them experience with instrument panels and flying in different wind conditions. Initially, the rate was $32.70 per hour for 180 hours in the 20-week training course. But British fighter aircraft such as the Supermarine Spitfire, an interceptor with elliptical wings, and Hawker Typhoon, a bomber for nighttime intrusions, flew at speeds exceeding 400 miles per hour. These new speeds posed new demands. In response, the Royal Air Force increased the training duration to 28 weeks.

While a Texas fog was no London fog, for aircraft, few conditions were more of a recipe for disaster than bad weather. Terrell was a "dust bowl when dry and a glue pot when wet," notes Killebrew. Flying in poor visibility was dangerous and even deadly. The weather often delayed commercial and cargo flights, and anyone worried about schedules was better off taking the train. Aviation couldn't become reliable until pilots were trained to achieve competency and confidence in any weather. As we'll see, overcoming these barriers required pilots to switch allegiance from their imperfect instincts to the aircraft's instruments, even when the latter contradicted their senses.

Training meant costs—in dollars and lives. Students named the Stearman, once famous as a crop duster, the "yellow peril." The fabric-covered Boeing biplane roared across the skies at 225 horsepower. Its open cockpit permitted a student and an instructor to sit in tandem. The Stearman was notorious for its "ground loop";

it circled out of control when landed improperly. Once, a student landed his Stearman on top of another, waiting on the tarmac for takeoff. In other cases, students who misused the controls overshot and crash-landed on the highway, in someone's backyard, or on trees. Some engines caught fire after landing, and some quit midair.

The wider aviation industry was shaking off turbulent experiences from the 1920s and 1930s and veering into a bold new age. Some years earlier, when the Navy and War Departments asked for a budget increase, President Calvin Coolidge had quipped: "Why don't they buy one airplane and let the aviators take turns flying?" By the 1940s, aircraft equipment was expensive and in demand. Inadequate maps and training manuals also accompanied the equipment shortage. The only way to learn how to fly at night and in bad weather was to fly at night and in bad weather. Still, a wartime footing demanded standards for professional training and proficiency. Aviation needed an approach that blended technology, psychology, and policy considerations and was realistic for initial and ongoing training.

The centerpiece of this lesser-known episode of the American and British military relationship was a general aviation trainer, the creation of Edwin Albert Link Jr., the chief protagonist of this book. Ed Link's pilot trainer consisted of a fuselage with mock wings and tail to give the impression of an airplane cockpit. Its accordion-like base bounced to imitate actual airborne motions, and its box circled 360 degrees, allowing infinite flight-like repetitions without injury. Under the hood were controls that replicated the cockpit experience.

Along with the stick and rudder pedals, an instrument panel with a clock and compass contained indicators for airspeed, altitude, turn and bank, oil pressure, and temperature. The pilots in this dark and cramped chamber had no outside horizon to reference, a precondition to train for instrument flight. The instructor sat at a desk with a map that traced the student's course using a special electrical recorder; the trainee and instructor communicated through two-way radio. The instructor sent electrical signals to the training unit at the turn of a switch. The trainer twisted and turned in a geometric ballet—puffing, wheezing, gasping, and groaning. The entire setup looked like a penny arcade ride. It didn't merely look like one; at one point, it *was* one. People dubbed it the Donald Duck, a winged armchair, and more memorably, the "blue box."

Ed Link asserted a systems approach for flight training. Even more significantly, though, his engineering brought women closer to aviation at a time when they experienced the barbs of discrimination. As more men enlisted for the war, the United States and United Kingdom faced a dwindling supply of instructors. Women became a natural fit as Link Trainer instructors. They initially worked as maintenance clerks at the Terrell flight school, fitting fabrics for planes and parachutes. Women were hired as instructors only after they agreed that they wouldn't date any cadets. Feelings, they were told, didn't have a place in an all-too-formal environment of "Mr." and "Miss," especially when women had to judge men's instrument readiness. The Link Trainer ultimately forged over half a million Allied Force pilots. With its synthetic environment, the Link Trainer didn't precisely portray the complexity of flying to the pilot but potently *implied* it. The Link Trainer was no illusion machine; it was an initiation.

"IN A DARK TIME, the eye begins to see," the poet Theodore Roethke wrote.

Ed Link's phone rang on the evening of February 10, 1934, and it was Charles "Casey" Jones, the world's airspeed record holder in 1931. The ace pilot, once taught by Wilbur Wright, was now the marketing man for Ed Link. Jones asked Ed Link to fly into Newark to demonstrate the pilot trainer to some army top brass. Two days earlier, President Roosevelt had ordered the Army Air Corps to take over mail delivery in the United States.

The morning of February 11 was drowsy and dark. In his leather helmet and goggles, the lanky 30-year-old Ed Link revved his Cessna and barked off the rain-soaked runway near Binghamton, New York. Frigid air whipped the windshield, and Ed Link was cold. He racketed over the Pennsylvania countryside, flying just above the tree level of the Endless Mountains with the Susquehanna River far below. The visibility deteriorated as thicker clouds swaddled him, but Ed Link pushed on. He turned left toward the Poconos, flying below the gray cotton-candy clouds. With hands stiff on the controls and eyes fixed on the instrument panel, Ed Link pressed ahead with a faint drone over the Appalachian ridges molded by the meandering Delaware River. The more tolerable terrains of northern New Jersey appeared but no signs of New York City through the thick blankets of gray. He was late for his meeting.

In Newark, the army officers got fidgety even as Jones assured them of Ed Link's arrival. As they decided to leave, Ed Link's steel screamed out of the fog and landed on the runway in a precisely timed fashion. The army officials could barely see Ed Link in the mist, even as he waved his hands from the cockpit. The landing was all the demo the officers needed on how to fly in bad weather. Gaining confidence in Ed Link's training system, they alerted Washington. Soon came an order for six Link Trainers at $3,400 each.

Aviation was different from other modes of transportation. It was the first to pursue "perfect all-weather operations, at least in any organized way," notes scholar Erik Conway. Operators of cars and trucks, or even trains and ships, didn't need to be concerned about blind driving—when required, they could stop and wait the weather out. But airplanes didn't have that luxury when airborne. Pilots couldn't circle for hours until the weather cleared because fuel was, after all, limited. Under these circumstances, as we'll see, delivering airmail was a perilous pursuit.

For most pilots in the 1920s and early 1930s, "contact flying"—keeping track of landmarks and railroads while flying close to the ground—was the norm. In limited visibility—or "blind flight"—lacking reference to the horizon, the pilots were disoriented, with their senses utterly unreliable. Their instincts contradicted their instruments—they felt they were turning in one direction while doing the opposite. They spun and stalled with vertigo. Accidents in which weather was a crucial factor "were dismissed simply as 'pilot error,' without further explanation," noted General Jimmy Doolittle, a pioneer of instrument flying. "This was true, but the error was too often the pilots' refusal to believe their instruments instead of their senses."

Before the army's abrupt interest in pilot training, almost no one was interested in Ed Link's flight trainer, and there was no market for it. As a teenager, Ed Link had spent a fortune on flying lessons. He thought that the overall expense—and the time—could be reduced if the bulk of the training could be done on the ground. Ed Link's hand-built invention, using the bellows, valves, and actuators of organs and pianos, coincided with the rise of instrument flying and enabled reliable and routine landings. A principal benefit of Ed Link's approach was that if people learned how to fly airplanes without a real airplane, then the real airplanes could be freed up for profitable use. His planeless instruction led to a psychological insight that not everything in the airplane had to be duplicated, only core parameters that gave pilots the essential flight experience.

The Link Trainer didn't merely apply an instrumental attitude to piloting. It generated a new coherence between those instruments and the person flying. That coherence grew beyond the confines of the trainer; it extended into the pilot's mind. This "synthesis of senses" has been compared to preparing a meal. Chefs can't decide whether a dish is ready through sight and smell alone and often use taste to confirm. Similarly, pilots employed sight, touch, sound, and even smell to understand flight conditions and guide their response. Now they added feedback from their instruments.

Ed Link's consequential landing in Newark is virtually unknown to the world. Soon after, more people discovered that airborne flight-training time could be slashed in half or more by simply learning how to fly on the ground. A central revelation in the story of the Link Trainer was, as one airmail pioneer paradoxically put it: "An airway exists on the ground, not in the air." That is, the development of ground-based training and organizations that support it was as crucial as the invention of the airplane itself. Ground-based training is now the default approach for cost and safety reasons. The Federal Aviation Administration requires all commercial pilots to gain their training for qualifications in flight simulators. And without simulators to train astronauts, there wouldn't be a space program. One observer wrote that it's "doubtful that any piece of equipment has been cursed more than the ego-busting Link Trainer" in the annals of aviation. The sweat of World War II pilots within the bounds of the blue box "would probably have floated an aircraft carrier." It's one thing to mass-produce 10,000 planes, but producing 10,000 qualified pilots to operate them in different conditions was an altogether different problem.

If the Wright brothers generated a "passion for wings," making airplanes a cultural symbol, Ed Link presented a principled protocol whose prime purpose was the preparation. Flight training proved as valuable as flight itself. In this book, we'll use Ed Link's work as a substrate to explore the cultural consequence of a systems engineering approach that should be recognized in the same league as the Wright brothers. Chapters are named after the six degrees of freedom: pitch, roll, yaw, surge, sway, and heave—that are an homage to Ed Link's engineering and the world it made. They relate the flight trainer's development from carnival attraction to today's immersive environments.

Ed Link was a person in multiple mediums—land, space, and water. He connected them with his boundary-blurring belief that good problem formulation was a prerequisite to success. Ed Link seamlessly crossed over problem sets unhindered by specialty credentials, like a cubist ushering art into its modern era. Ed Link was thoroughly conversant in technology, behavior, and policy, what his device trained, and how it ordered and organized the experience. That's why ground-based training has resonated well beyond aviation and can now offer value to engineering our wicked problems in business, health, education, and policy. The confinement in Ed Link's blue box gave trainees the consciousness of the many angles to approaching a problem. In a way, it's an invitation to think *inside* the box.

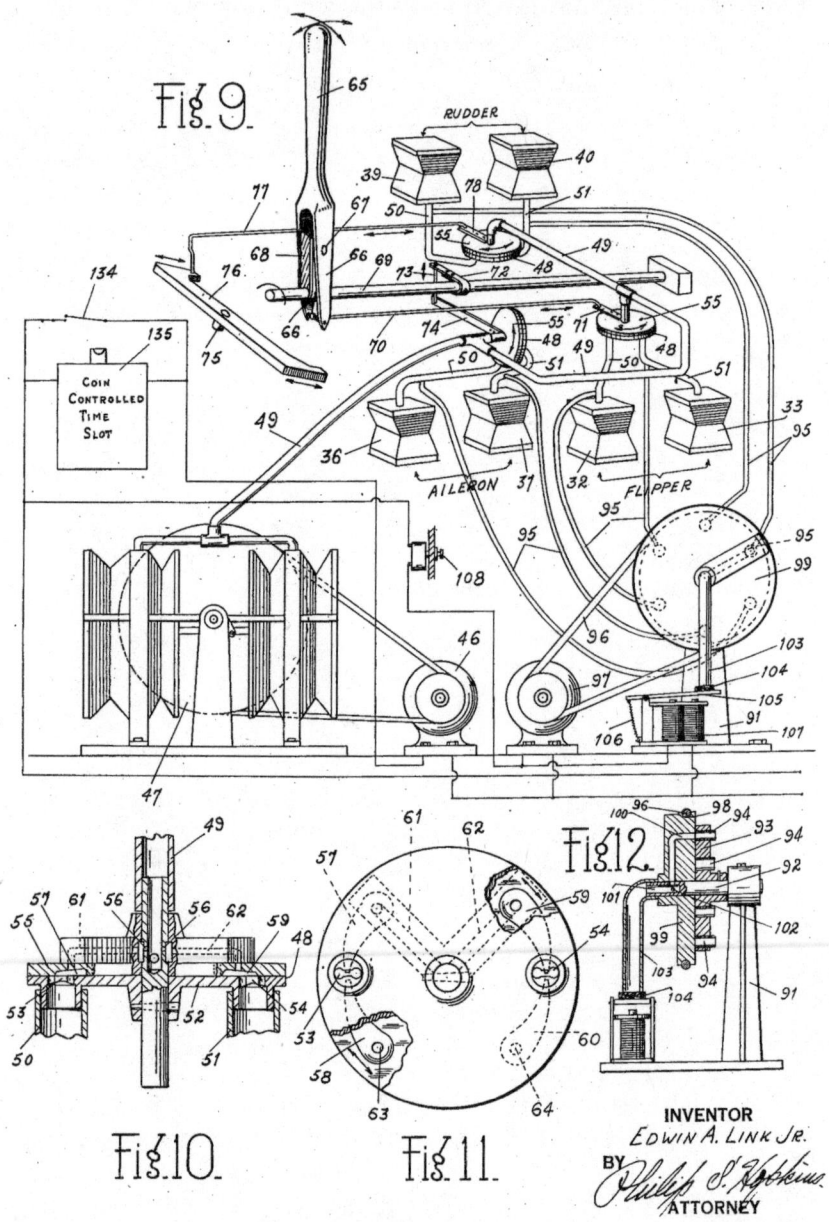

Fig.9.

Fig.10.

Fig.11.

Fig.12.

INVENTOR
EDWIN A. LINK JR.
BY
Philip S. Hopkins
ATTORNEY

Chapter 1

Pitch

I N AUGUST 1927, LIEUTENANT NORMAN GODDARD WAS hell-bent for Hawaii. So much so that he painted the letters H.B.H. on the rudder of *El Encanto*, the glistening, all-metal monoplane favored to win the open-ocean race from California to Hawaii.

Earlier that May, Charles "Slim" Lindbergh had soloed his 450-gallon, single-engine *Spirit of St. Louis* nonstop for 33½ hours from New York to Paris. Until then an unknown airmail pilot, Lindbergh's fame soared as he clinched the $25,000 purse put up by hotelier Raymond Orteig. James Dole, a charismatic Harvard-educated farmer turned pineapple magnate, saw an opportunity here. By hosting a flight contest of his own, Dole wanted to produce the Pacific Lindbergh. The first prize for the nonstop feat was $25,000; the second, $10,000.

Soon after the contest announcement, two army pilots pioneered the transpacific flight in a plywood-skinned Fokker Trimotor, *Bird of Paradise*. Favoring fuel more than food in payload, on June 28, 1927, Lester Maitland and Albert Hegenberger winged their way from the San Francisco Bay into the broad blue expanse. After 25 hours and 50 minutes, they landed in paradise. However, the civilian glory to end Hawaii's isolation was still up for grabs. The *San Francisco Examiner* called the 2,400-mile contest the "greatest air race" in world history, comparing its life-and-death spectacle to Roman gladiators entering the arena. Fifteen competitors signed up. But tragic crashes the week before the derby shrank the final race list to eight contenders. A ninth entrant, *Spirit of Peoria*, remained a possible contestant. Its pilot planned to carry a caged owl christened Colonel Pineapple to stare at him and his navigator to keep them awake. But the biplane was disqualified for inadequate fuel capacity.

On Tuesday, August 16, Dole's race day, Goddard and his navigator, Kenneth Hawkins, arrived in suit-and-tie, calf socks over pants, to set off to Honolulu from Oakland. They tested their radio adapted from high-end navy equipment, the detachable landing gear, and the inflatable rubber wing float. "When we get over to Honolulu we're going to reserve rooms for the rest of the racers," Hawkins said. "After that I'm going to rent a surf board and an outrigger canoe and practice up so I can show the boys a few tricks when they arrive." After days of preparation and photo ops, *El Encanto* was ready, with Wright Whirlwind engine, salmon-like fuselage, and three gas tanks, topped at 350 gallons.

The daredevil racers and nearly 100,000 enthusiasts who took the day off waited for a fog bank to clear at the Oakland field. Spectators showed up with their picnic chairs, snacks, and ukeleles. Those in San Francisco thronged every hill and high ground with binoculars. Around noon, the starter waved the checkered flag—a sharp silence set in.

First up, blue-and-yellow *Oklahoma* thundered down the runway and lifted off to a loud shout. Two minutes later, the crowd cheered anew as Goddard throttled *El Encanto*. The silver machine roared toward a 120-mile-per-hour takeoff but clumsily swerved left and right. The plane's wheels caved and toppled in smoke and dust, skidding Goddard's hopes and dreams 4,800 feet from the start line. Sirens wailed, and fire trucks rushed to the wreck; the announcers were relieved that the gas tanks didn't blow up. Goddard and Hawkins scrambled out, swearing at their plane. "I would rather have crashed in midocean than to have had this happen," Hawkins said.

Next up, *Pabco Pacific Flyer* rumbled, rattling into the nearby marshland, 7,000 feet away. The pilot and navigator, Livingston "Lone Eagle" Irving, crawled out in rage; a tractor towed his shining orange plane back to the field. Then, at 12:30 p.m., the sleek, cigar-shaped Lockheed Vega *Golden Eagle* whizzed off for the journey west as Goddard and Hawkins kept bickering over whose mistake it was. A minute later, the red-white-and-blue biplane *Miss Doran*, flown by a circus pilot accompanied by the aircraft's namesake, a 22-year-old Michigan schoolteacher, barreled down the runway. It barely rose, and circled back. The remaining three contestants—a cadmium-colored *Aloha* piloted by Martin Jensen, the yellow-blue *Woolaroc* with the Hollywood stunt flier Art Goebel, and the silver-green *Dallas Spirit* flown by the combat veteran Bill Erwin—cleared the runway.

But within minutes, *Dallas Spirit* returned with its fabric fuselage ripped loose. Then, the third plane to return, *Oklahoma*, hovered low over the crowd with a trail of dark smoke, strangled by a malfunctioned motor. *Miss Doran* sprinted off swiftly on its second attempt. "There it is; there's the Pacific Ocean. We're on our way for sure this time," said Mildred Doran, the young teacher on board. *Pabco* was ready for another try. It bounced uncontrollably for a slow climb, toppling over its wing. "Let's go home, you've had enough for one day," said Irving's wife. The crowd started to drift away, betting on the four planes that took off in the grand contest, leaving the rest behind, disabled and damaged.

Woolaroc steadily hummed dots and dashes after a day of flying. Art Goebel grew worried that he should have seen the islands by now. He was almost deaf with the continuous engine noise. His navigator, William Davis, assured him they were on course and Hawaii was within three hours. Goebel panicked, as the remaining fuel would last only two hours. Water landing seemed inevitable. But Davis boosted his radio range to develop a better fix and double-checked his compass. Soon they spotted a dim shore in the distance and erupted in joy. "Here we are," Goebel said. "By gosh, my job's done." Goebel and Davis started shooting pistols and dropping smoke bombs above Koko Crater and Diamond Head. *Woolaroc* reached Honolulu in 26 hours and 17 minutes, winning the first prize from the Pineapple King.

Two hours behind, Martin Jensen and his navigator, Paul Schluter, couldn't avoid the clouds. They flew *Aloha* a hundred feet above the water for many hours, eventually spotting Oahu. "For God's sake don't do any stunts," Schluter told Jensen. "We are lucky to be here." *Aloha* landed second as crowds screamed louder. Hours later, Dole grew agitated waiting for the other two planes yet to arrive: *Miss Doran* and *Golden Eagle*. And they would never come. Three days later, the mended *Dallas Spirit* flew a rescue mission with a salvaged radio from the wrecked *Pabco Pacific Flyer* to search for the two missing planes. "I believe with my whole heart that we will make it. . . . We will win because Dallas Spirit always wins," the pilot Bill Erwin wrote in his departure note. "It is the last and most wonderful adventure of life." The *Dallas Spirit*, too, disappeared. One poet wrote:

> The days crawl by—no hopeful words
> Are wafted back of the lost Dole-birds,
> Tell us, oh Gods of the Air and Sea,
> Where are the seven who soared so free
> Into the West?

The Dole Derby was a Pyrrhic pursuit, with two victors and 10 victims. Newspapers declared it "aviation asininity" and an "orgy of reckless sacrifice." Even *Aloha*'s pilot, Jensen, said, "The expenses turned out to be a lot higher than the winnings." Success in this mission required velocity and vision, reliability and recovery, fuel and faith. But the race fundamentally exemplified the danger of blind flight, in which pilots flew through treacherous conditions without the know-how to navigate clouds, fog, and poor visibility. These fliers needed the skills to interpret and trust the instruments that enabled them to gauge their relationship with the horizon. They often stalled, spiraled, or spun out when disoriented and dizzy. *Aloha*'s navigator, Schluter, vowed never to fly to Hawaii again. The winning Art Goebel, who was the first to sign up for the Dole race and also dreamed of spiraling around the Eiffel Tower and the Brooklyn Bridge, gave up stunt flying. Jensen was on a job with a movie studio to fly its mascot, a lion, across the country when he crashed his plane in poor visibility over the Arizona wilderness. Lost in the desert with a 400-pound cat, Jensen trekked for three days until he got to a ranch to call the studio, which was more interested in the lion.

"America has found her wings," Lindbergh, who declined to participate in the Dole race, wrote, "but she has yet to learn to use them." It took a few more years—and far more lives—before pilots had access to technology that could teach them flight skills without risking a crash. This technology would eventually take shape in Binghamton, a city in upstate New York with an unexpected legacy of engineering.

THE BINGHAMTON REGION WAS a vital trading post in the 17th and 18th centuries. During the American Revolution, the land belonged to the Haudenosaunee, whose name means to "form a cabin," connoting the strength and loyalty in the alliance of six nations of the Iroquois Confederacy. The crooked Susquehanna River and a calmer Chenango, blending in Binghamton, brought commerce, travel, and trade.

In 1786, a British barrister turned real estate mogul, William Bingham from Philadelphia, co-purchased about 33,000 acres of land at 12½ cents each. The owners divided the lot in 1790, and Bingham's 15,000 acres included today's Broome County. Ten years later, having never visited the area, Bingham, primarily working through an astute agent, Joshua Whitney, sold the lots for an extraordinary 10 to 15 dollars an acre.

In 1817, the engineering of the Erie Canal began. To some, the nearly $8 million, 363-mile project was the "Eighth Wonder of the World," and to others, it was the "Big Ditch." The towpath from Binghamton, a nearly 100-mile-long reservoir-fed waterway known as the Chenango Canal, opened in 1837. It had been hand-dug through mires and marshes by hundreds of migrant laborers.

In the years that followed, the lush rural economy became more extractive. Plow shops took root, and logging and rafting operations sprang up. Dairy farms swelled. The "iron horse" of the railroads led to fresher produce and faster businesses. Factory furnaces were commonplace in the 1850s. In the 1870s, some firefighters christened Binghamton "the Parlor City" as a homage to its hospitality that far surpassed their hometown, Scranton, Pennsylvania. Years later, a popular song celebrated Binghamton as "clean, fair, and grand."

IN 1898, THE FIRST coin- and roll-operated piano was launched in the United States by Peerless Piano Player Company of St. Johnsville, a small town a hundred miles northeast of Binghamton. Their products featured controls for automatic roll rewinding at the tune's end. Such machines played masterpieces on demand—one could loop Schubert's *Unfinished* Symphony no. 8 in B Minor, Chopin's *Heroic* Polonaise in A-flat Major, and Dvořák's *New World* Symphony in E Minor. But then came a central ambiguity, in the words of an attendee at a player piano concert: "Should one applaud? For nobody is there. It is only a machine."

Between 1910 and 1925, about 85 percent of pianos made in the United States were player pianos, or pianos that played themselves. Their popularity in the early Jazz Age created a lucrative industry rivaling real estate and automobile businesses. The pianos brought forth "every touch in technique, every subtlety of expression and tone color; in fact, you hear the actual playing of a master musician," one company claimed. In 1911, one Niagara Falls company advertised a console with a three-hour, four-course dinner selection. The opening course was accompanied by the harmonies of Richard Wagner, the second course with the tunes of Ignace Leybach, and the third with Albert Gumble and Giuseppe Verdi, and the dessert concluded with ragtime. The player pianos were pneumatic virtuosos, powered primarily by suction. The basic design came from Alexandre-François Debain, a French instrument maker who honed the harmonium in the 1840s, whose portable versions, common in the Indian subcontinent, require hand pumping. When pressed, the player piano's foot pedals controlled airflow into the bellows and through it to the machinery of tubes, gaskets, hoses, inside valves, outside valves, pallet valves, flap valves, slide valves, and crankshafts. The air ran through the pipes and produced the piano's tones.

The music was "programmed" on holes spaced across a spool of rolled paper that turned in one direction at the desired speed and, in turn, defined the tempo. The pianist's pedaling accented the melody. After 1908, player pianos used standardized music rolls better suited for complex classical scores. The 88-note rolls had 88 holes spaced at 9 holes per inch on a standard 11¼-inch paper width and ran up to 100 feet in length. The application of perforated holes for music notation came from textile manufacturing, courtesy of French weaver Joseph Marie Jacquard, who, in 1801, engineered a loom attachment to automate patterning. The "Jacquard loom" followed earlier innovations by the machinist Jacques de Vaucanson, who automated silk rolling using pierced metal cylinders and launched the lathe. This momentous machine spun off the Industrial Revolution, and the same concept later revolutionized punched cards for computing.

Now, with player pianos, picture a conveyor belt or, loosely, a cassette tape. The information stored on the roll was read as the paper (instead of a magnetic film) moved over a tracker bar with holes arranged in a row perpendicular to the roll. The opening activated the pneumatic apparatus when the tracker and paper holes aligned. Compressed air from the bellows traveled through the pipes and valves to raise or lower the piano keys. Some units, called orchestrions, came with even more elaborate arrangements. They were attached to a complete wind orchestra, mandolins, organ pipes, traps, timpani, and triangles to ring out a medley of classical and popular tunes. And some player pianos spontaneously played when one deposited a coin.

Some argued that "programmed" music machines alienated artists from the stage and from themselves. "The nightingale's song is delightful because the nightingale herself gives it forth," one critic noted, sternly resisting "the menace of mechanical music" in 1906. "Music teaches all that is beautiful in this world. Let us not hamper it with a machine that tells the story day by day, without variation, without soul, barren of the joy, the passion, the ardor that is the inheritance of man alone." But mechanization also brought benefits: "Automatic instruments were always ready to play, day and night. They never went on strike, never were late, did not take breaks, and never came to work drunk." Competition led to a manic mechanical music market and an industry in flux, frenzy, and fight over design tweaks, which scholar Brian Dolan characterized as "chronic mechanitis."

One Binghamton-based firm, the Automatic Musical Company, purchased high-end pianos from the Schaff Brothers Piano Company of Huntington, Indiana, and added wooden casing and glass frames to Schaff's products. The Automatic Musical Company improved on the self-playing mechanism by using "endless rolls" with superior rewinding and replay features. However, expenses for a bitter patent dispute with a competitor edged the company toward bankruptcy in 1913. So, George T. Link at Schaff Brothers dispatched his son Edwin from Indiana to Binghamton to take over and operate the Automatic Musical Company. Edwin Link Sr. eventually purchased Automatic and incorporated it as Link Piano Company in 1916.

At its high point during the "golden age" of automatic instruments, Link Piano Company had 125 employees. The firm focused on expanding the reach of coin-operated pianos and orchestrions, competing with Peerless and the "mighty Wurlitzers." They also made repeating xylophones and some units without keyboards. The endless rolls in the Link piano units—which now lasted up to 15 tune lengths—provided greater variety for silent pictures: songs to accompany romance and fighting, dancing and pathos, humor and tragedy. Their versatile units were "distinctly a delight" in producing special effects, from screaming spirits to galloping horses. They eliminated technical annoyances and ensured constant music speed as harps and flutes swiftly harmonized with the push of a button.

The Link Company's pianos weren't a success for just theaters. Their nickelodeons, or nickel-operated pianos, drove traffic—and business—at soda fountains, ice-cream parlors, corner drug stores, skating rinks, ballrooms, and cabarets. High-end clubs and restaurants sought quarter and half-dollar slots for the affluent clientele. By 1925, the Link Company produced about 300 nickelodeons, 12 pipe organs, and many music rolls each year. The firm even launched its first automatic theater organ. With three keyboards and 11 ranks of pipes, the Link C Sharp Minor organ's mechanical operations were virtually noise-free, even with its electrical motors and generators.

One leading New York organist, fittingly named C. Sharpe Minor, wrote in a 1925 letter that the instrument was the "most pleasant surprise. The quality of the tone was beautiful, and the response of the action superb. I had never known before such instantaneous response from any pipe organ." The Link organ, "tonally and mechanically, was far ahead of any organ on the market today," he observed. Installations of this "architectural symphony" with elaborate jewelry and velvet curtains spread nationwide, from Elk lodges to opulent residences and mausoleums. In 1927, the organ became the centerpiece of Binghamton's majestic Capital Theater, well known for its variety shows. However, the silent-era popularity of Link theater pianos and organs was short lived. With the rise of the "talkies" and the economic downturn of the Great Depression, player pianos and organs were relegated to cultural obscurity.

But the technologies within the Link theater pianos would not remain backstage for long. Ed Link's inventiveness soon found them an unexpected second life in aviation.

HISTORIAN JOSEPH CORN DESCRIBED America's affection for aviation in the early 20th century as the "winged gospel." For some Americans, an airplane was a mechanical god, and their pilots, divine heroes. Aviation symbolized the promise of a peaceful future in the heavens. Like traditional faiths, aviation imbued people's lives with meaning. And like secular social visions, aviation became a technological reform movement. The resulting "air-mindedness" evoked utopian images of prosperity where each person could one day own an airplane. In Europe, Pablo Picasso oil-painted *Our Future Is in the Air* on canvas for a 1912 pamphlet promoting aviation. Picasso and Georges Braque compared themselves to the Wright brothers since their cubist creations were similarly trailblazing. The same year, the French dramatist Romain Rolland wrote: "Take possession of the air, submit the elements, penetrate the last redoubts of nature, make space retreat, make death retreat."

Despite the dangers of aviation, many pilots contributed to the romanticized notions of flight. In a 1924 issue of *Cosmopolitan*, the one-time pilot Winston Churchill, who had survived a take-off crash, wrote that "air is an extremely dangerous, jealous, and exacting mistress. Once under the spell, most lovers are faithful to the end, which is not always old age." Even the safety features of 1920s aviation seemed more like fashion for a debonair bachelor than protection. "Goggles, gloves, a heavy leather coat, and a cork-and-leather helmet were de rigueur," writes scholar Peter Pigott, and the Wright brothers dressed in their Sunday best to fly.

But, as the Dole Derby and other disasters demonstrated, flying was highly dangerous. Pilots faced enormous demands with maintaining complex coordination between what they saw clearly when they could and how they felt about it through their internal sensations of ears, nerves, and muscles. Before radio and other navigation aids, airmail pilots relied on tracking what was visible on the ground. In this "contact flying approach," fliers focused on a familiar landmark. They flew toward it, then fixed their sights on another landmark to track, repeating the process until they reached their destination. The transcontinental air system grew from the Union Pacific Railroad route to San Francisco by siting 15 landing fields, spaced 200 miles apart. Decades before satellite navigation, airmail pilots used Rand McNally road maps they could pick up at gas stations or the "iron compass" of the railroads, water towers with painted town names, and lonely ranch houses.

Pilots memorized the heights of hills, steeples, water tanks, and telephone poles so they didn't crash into them. And if pilots flew long distances, it was often at low altitudes, about 50 to 300 feet from the ground, always ready for emergency landings. Occasionally, aircraft got snarled in power lines, as was the case with a Canadian pilot in 1918 who was suspended above a store building for hours. One pilot said he got around "with a few drops of homing pigeon" in his veins. When pilots needed timely wind data over the farms, they looked to the cows below as a proxy since the animals always grazed with the wind behind them. If the bovine compass failed, they looked down for direction hints from rural outhouses, which always faced south. While this method may have worked in local landscapes, contact flying over enemy territories was extremely unsafe. The challenges of wartime flight demanded better pilot skills and higher flight.

In 1920, the US Army began a survey of "sky roadways" to improve airmail service. They mapped 44,000 miles with data from 90 flights. This effort resulted in a comprehensive listing of relevant landmarks, rooftop markings, and possible landing fields and patches, where excited crowds sometimes interfered with aircraft landing. A 1921 follow-up survey provided a log of coast-to-coast distances, landmarks, and flying directions for a transcontinental air network. The routes stretched from New York City to Bellefonte, Pennsylvania, then to Cleveland, then to Chicago, then to Omaha, and the longest leg to Cheyenne, then to Salt Lake City, then to Reno, and finally the shortest leg to San Francisco. Pilots preempted modern-day voice-guided navigation by writing down almost turn-by-turn directions every few miles. Their notes pointed out lakes, trees, railroads, telephone cables, and hangars.

One popular survey book, *Landing Field Guide and Pilot's Log Book*, often contained colorful commentary: "Following Michigan Central double track railway . . . Good fields but country is getting more rolling." "Plenty of fields large enough but one would have to be very careful to avoid granite boulders which are there in large numbers and sizes." "Absolute wilderness . . . Still flying compass course. Have not seen landmark house, field, town or sign of human life for one hour and ten minute until just now."

The book also contained helpful hints. "Don't fly if you can't walk straight; Don't invent new stunts; Don't be a flying fool a damn fool is safer." And then some caution: "Be conservative . . . Feel your Machine; don't let it fly you . . . It is better to be a Sky-Pilot than a Crack-Ace. Be conservative." And some bonus instructions: Don't drop ballast other than fine sand or water from the aircraft. Should the engine fail on takeoff, land straight ahead no matter the obstacle. A plane with a lifeless motor had the landing right-of-way, and no aircraft was to taxi faster than humans walking.

To ensure the mail crews were oriented correctly, in 1923, the US government created the "lighted airway" by drawing hundreds of yellow concrete arrows, each about 70 feet long, for fair-weather guidance. Each arrow had an adjacent 50-foot rotating acetylene searchlight to guide night flights. A single beacon—at 500 watts with 18-inch doublet lenses—provided 10 miles of visibility under clear conditions and automatically turned off during the day. About 250 of these beacons illuminated the Chicago-to-Cheyenne routes.

In 1924, the US cross-country flight service was born due

to lighted airways. The New York–to–San Francisco route had 10 radio stations, and the US Navy dispatched weekly weather reports. In these early days of cross-country flight service, a trip from east to west took about 34 hours, and the reverse, 29 hours. Subsequently, NOTAM, or Notice to Airmen (now called Notice to Air Missions), sent regular hazard updates along flight routes, brimming with codes and acronyms. By 1925, there were about 100 commercial planes in service. The following year, with approval from Congress, the postmaster general contracted out mail delivery to private companies. As we'll see, this move created a spate of complications, including a political debacle.

As cross-country flight became increasingly popular, one daredevil acrobat and airmail pilot, Elrey Borge Jeppesen, realized the need for standardized pilot directions. Bad weather could blur even the bright beacons. "Sometimes you couldn't see a mile, so what the hell good was a beacon?" Jeppesen said. So he purchased "little black books" at 10 cents each to start compiling directions to amend and improve on government aerial charts. While piloting aerial survey planes and photographing terrains, Jeppesen made notes of his surroundings that would prove lifesaving to other pilots. Jeppesen bought an altimeter to estimate the heights of landmarks—from mountains to smokestacks. He approximated safe landing distances for emergency strips and scouted alternative travel routes. His notes contained practical details such as landing locations, runway lengths, and simplified terrain views and altitudes.

Jeppesen included information on where to get the best weather report and fuel and what to do when successive storms gripped pilots. "He knew, for example, that when approaching a certain airfield, he had to keep the grain silo to his right and the road with its line of telephone poles to his left," Jeppesen's biographers note. "Or, when taking off from another field, he had to pull up enough to clear the barn and windmill and grove of trees at the end of the runway. Simple enough, but what about when the weather was bad or the visibility restricted? Exactly how far from the runway are those poles, that barn, and those trees? Just as important, how tall were they?" Jeppesen's scribbles and sketches—which he developed into standardized flight maps and called the Jepp charts—became a sought-after resource for pilots. "I didn't start out to chart the skies," Jeppesen said. "I did it to keep myself from getting killed."

By 1930, five major airlines existed, together carrying 400,000 people per year. As air travel became more popular, air-traffic problems soared. To ensure that planes didn't try to land simultaneously or fly toward each other at the same altitude, eastbound aircraft began flying at odd-thousand-foot altitudes and westbound flights at even altitudes. And while in 1923 only 250 beacons helped guide pilots, by 1933, 1,500 beacons lit American airways. Pilots relied on Morse-code dots and dashes in their cockpit AM radio receivers to differentiate the beacon lights they tracked to their destinations. They were flying on course if they heard a steady tone; otherwise, they were off course and needed to correct. Further, the beacons relayed their course with single Morse letter flashes through the dark: 1 for W, 2 for U, 3 for V, 4 for H, 5 for R, 6 for K, 7 for D, 8 for B, 9 for G, 10 for M. The letters formed the mnemonic "When Undertaking Very Hard Work Keep Direction by Good Methods."

As we'll see, a high-profile crash involving a US senator in 1935 goaded the government to create a national air-traffic management system, which fully emerged only in the early 1940s. But until then, the problem of blind flying remained, and lousy weather cut short journeys and lives. The successful solution, however, required Ed Link's ingenious engineering—to make pianos fly.

BORN IN 1904, ED LINK was mechanical minded. Eager for solitude, young Ed found his fun deconstructing and reassembling clocks. His other pastimes included tinkering with gunpowder at a friend's farm and blowing the door off the barn. There were other projects: building toy cannons that shot kerosene bombs, assembling a mechanical printer to produce a comic newspaper he sold for a penny, and designing an inflatable slide from his bedroom window to the street. Whenever he walked down the school hallway and entered the classroom, his jingling pants pockets, bulging with keys, chains, and knives, drew attention with every step.

At 12, Ed Link drew a pencil sketch of a submarine. "Unfortunately, it wasn't very practical; it most assuredly would have sunk," he confessed years later. His technical aptitude aside, Link was an academic disappointment. Ed Link was "just dumb enough to be a genius," an associate recalled. He was expelled from classes and dropped out of junior high school. Later in life, Ed Link maintained that formal education taught all one couldn't do.

In 1920, when living in California with his mother, Katherine, who was by then separated from Edwin Link Sr., young Ed Link caught the "air-mindedness" that acted on the imaginations of adventurous kids. During Ed Link's youth, many out-of-work pilots who had flown in World War I became acrobatic showmen. These nomadic barnstormers traveled county to county and coast to coast, stopping in cow pastures and at state fairs to entertain the locals. They organized flying circuses and showed off inversions, loops, and snap rolls at 100 miles per hour. These bold "birdmen" charged spectators 5 or 10 dollars for a few minutes' ride in their Curtiss Jennys—wood-framed surplus war stock described as "a bunch of parts flying in formation." The Jennys tended to spit oil and fumes on the pilot's face. They occasionally caused a fire in the plane's flammable fabric. Daredevil solo fliers walked on wings, changed planes in motion, and even played tennis while aloft. Some vaulted from aircraft to running boats or trains.

One morning in 1920, young Ed Link borrowed a motorcycle to drive to an airfield on Wilshire Boulevard in Los Angeles to "Fly with the Stars." A year earlier, Sydney Chaplin—Charlie Chaplin's brother—became a proprietor of a short-lived airline company. Ed Link had saved money from his work repairing motorcycles. He signed up for three lessons from Chaplin for $50 an hour (close to $750 an hour in today's value). In aviator gear replete with a silk scarf, Chaplin offered Ed Link a seat in the back of his Curtiss Oriole. In a death-defying experience, Ed Link got a lightning tour punctuated with loops and spins, where he hardly touched the controls. "That's a hell of a way to teach someone to fly," Ed Link later recalled, pointing out that most "old-time aviators, like Chaplin, started teaching their students by scaring them half to death."

Ed Link's family initially disapproved of his interest in aviation. When he returned to Binghamton in 1922, after short-lived jobs in Illinois and West Virginia, Ed Link joined his father's piano and organ firm. A grease-under-fingernail tinkerer, Ed Link revived and restored decrepit player units as his brother George oversaw the company's operations. In 1924, Ed Link rigged a vacuum cleaner–style device to remove lint from the piano tracker bars, which degraded the endless-music rolls. This fix solved a frequent nuisance that made the units dysfunctional. A year later, he directed the design and rollout of the Link C Sharp Minor organ, which the company installed in the hundreds. Undeterred by his parents' disapproval, in 1926, Ed Link completed his first solo flight, with lessons from a local aviator. Edwin Link Sr. was displeased with his son's passion for flying and fired Ed Link from the factory in response, but he quickly rehired him. Ed Link's inventive instincts clicked into gear during his lessons, perhaps spurred by the necessity of his penny-pinching to afford flight school. He began to imagine new ways to slash the high costs of pilot training. His idea sparked a new era.

WILBUR WRIGHT TRAINED HIS pilots to gauge the effects of wind by opening and closing the hangar doors. He sat at the controls and told his students to "imagine" air disturbances and the necessary aerodynamic response as they maneuvered the parked plane's rudder and wing controls. The idea of "flightless flight" defined many inaugural efforts for flight training.

One contemporary outlier of the Wrights was Glenn Curtiss, a motorbiking pioneer from Hammondsport, a Finger Lakes village 100 miles west of Binghamton, who embraced a different method. Because his planes were all single-seaters, there was no way students could learn from their instructor on the flight. Learning from others' airborne demonstrations and grading the flight from the ground was the only way. Under Curtiss's instruction, students translated what they observed from the ground into flight maneuvers. In contrast, the Wrights emphasized the "aerial" experience the students cultivated on the ground.

In 1910, the Wrights developed a "kiwi-trainer" in Huffman Prairie, Ohio, named after the flightless bird. In this trainer, students sat in a Wright Model B pusher biplane mounted on a motor-driven platform. They were introduced to basic flight movements and sensations. The Wrights charged an extraordinary one dollar a minute for these elementary lessons (approaching $2,000 an hour in today's value), and their courses were usually four hours minimum.

Around the same time, engineers developed several other "motion trainer" concepts. A British automobile engineer, Eardley Billing, came up with the idea of an oscillator. This rudimentary biplane sat on a rotatable undercarriage. The wobbling gave students a "feel" for open-air flight. At the Antoinette Flying School grounds, a French military camp, students learned simple balancing maneuvers while seated in a wooden barrel split lengthwise. When instructors disturbed the ropes in the mounting arrangement, as in tug-of-war, the student had to control the barrel's action and align with the horizon by adjusting pulleys.

In a similar effort to reconstruct balancing motions experienced in flight, after crashing his biplane on a telephone pole, British captain Haydn Sanders recovered the wreck, stripped out the motor, and sat the plane on a metal pivot. As the contraption rocked, the students sitting in it learned to operate the rickety actions of the elevator, rudder, and ailerons. But this device failed because it required steady wind.

World War I brought on new attempts to design flight trainers.

One wacky-seeming design came from New Yorker William Guy Ruggles. He affixed a chair in a gimbal-ring assembly powered by electric motors. The unit spun on all three axes, creating comic disorientation and nausea in students. Its inventor was confident that with blindfolding, one experienced intense illusions of air-sickness. Ultimately the Ruggles Orientator—or the "bathtub," as some called it—only assessed a candidate's fitness to fly rather than training them how to fly.

To further improve on the Ruggles Orientator, one visually impaired Philadelphia native, Colonel William Charles Ocker—later called the "Father of Instrument Flying"—proposed a hooded version to create the nonvisual conditions. With his partner, Colonel Carl Crane, Ocker taught aviators to rely on instruments when they could not see the earth's surface. Ocker and Crane's conception of blind flight didn't mean flying blindly. In darkness, fog, and other poor-visibility conditions, "some artificial means must be used in order that flight may be possible," they wrote in their influential book, *Blind Flight in Theory and Practice*. They argued that one could control an airplane without a proper visual reference through instruments. But blind flight entailed two interlinked problems: one, flying the aircraft without external visual reference, and two, controlling the aircraft without outside vision. Again, these two problems demanded that pilots learn to interpret and trust the instruments. Granted, this was not a simple task; other efforts continued to provide more realistic training. The French Air Service used old Blériot monoplanes with a truncated wing-span and had trainees taxi at 40 miles per hour. They called this use of a reduced-wingspan plane the "penguin system." The American version of this method, using stub-winged Curtiss Jennys, was called the "grasscutter." In both versions, the aircraft didn't take off, so the students could merely get acquainted with the basics of flight dynamics.

While working in the piano factory, Ed Link stayed active at his local airfield, rehearsing his barnstorming skills, including barrel rolls and parachuting stunts. He was ecstatic about Lindbergh's transatlantic achievement in the summer of 1927. He learned about the Sanders Teacher and the penguin system. But it was Ocker's work that compelled Ed Link to initiate his own training system. Ocker had recently blindfolded pilots during training, turned them around on a swivel chair, and asked whether they turned clockwise or counterclockwise. "And that was one of the things that gave me the idea that you could make a whole airplane to train a pilot to do everything," Ed Link recalled for an oral history project five decades later. "He was merely demonstrating just what I repeated: that you couldn't tell where you were going by sight or feel. You had to have an instrument that told you were turning and whether you were flying straight or level and so forth." But with the ground-based trainers Ed Link envisioned, pilot training could happen readily at scale. "They just thought that 'twas silly, but time has proved differently."

As the engineer turned sculptor Alexander Calder—heralded as the painter of movement—once said, "Just as one can compose colors, or forms, so one can compose motions." In late 1927, Ed Link began engineering his motion trainer—part piano and part plane.

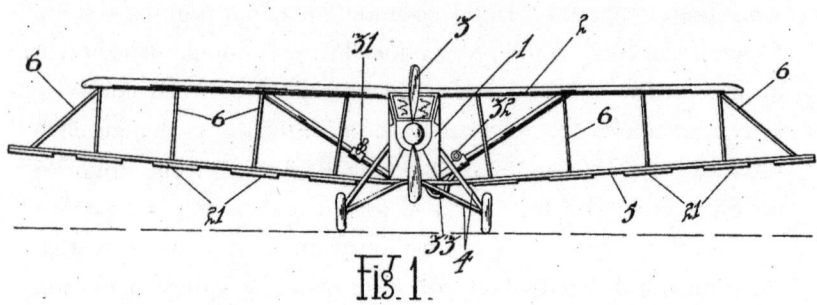

Fig.1.

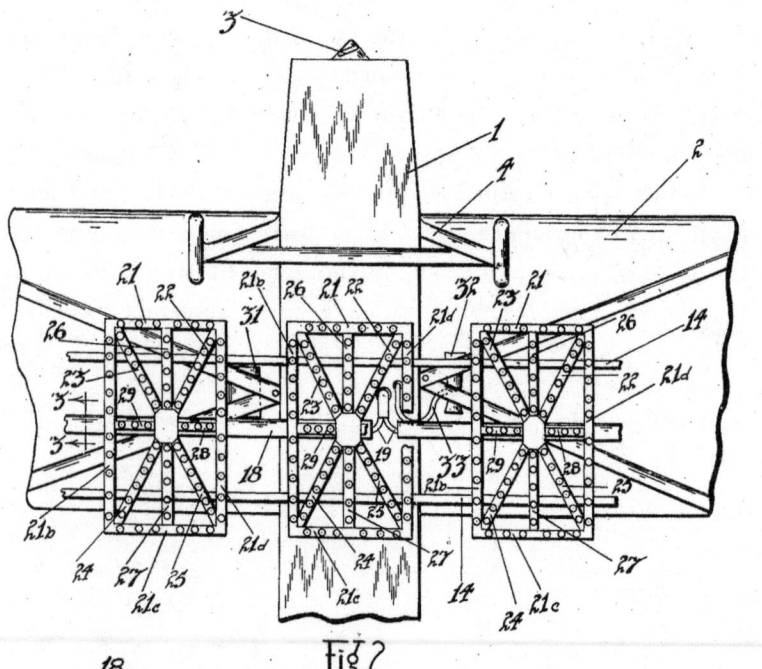

Fig.2

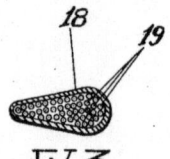

Fig.3

Chapter 2

Roll

O RVILLE AND WILBUR WRIGHT BEGAN PURSUING AERO-
nautics as a sport and reluctantly entered the scientific side of
it. Twenty years later, Ed Link experienced the opposite effect with
his flight trainer. He was reluctant to see his engineering become
entertainment, but he wasn't surprised. As his patent claimed, the
Link Trainer provided "great value as means of instruction for student
pilots" and "affords an interesting and unique entertainment device."

Ed Link's generic aviation trainer adapted the suction principles he'd used to great success in player pianos. But instead of
making piano keys go up and down, the electrically activated bellows drove compressed air that mimicked the aileron, elevator, and
rudder functions. Like the piano pedals, the stick movement filled
or emptied the bellows to duplicate pitching and banking motions.
The bombinating device flew and felt like an airplane. The pilot
trainer's initial version featured a hooded cockpit mounted on a
turntable, a turn-and-bank gauge, a compass, and indicators for
airspeed and rate of climb. A universal joint provided the connection between rotating shafts. The machine's overall setup looked
like an eight-foot bumblebee turning in a 15-foot diameter.

In 1930, Ed Link opened a flying school featuring his unique
invention. He charged a flat $85 fee for the course. In it, pilots
safely rehearsed at a fraction of actual flying costs without concern
for losing life or equipment. The intercom allowed the student and
the supervisor to communicate during the session. The instructor's

desk also contained an automatic plotter of the student's course. Driven by a motor and three wheels, the crab-like record keeper allowed the students to review and correct their mistakes in subsequent training sessions. In the ensuing years, the instructors, often not pilots themselves, proved as vital as the instruments.

The flight was defined by "physical movements and sensations, the signals and indications from the cockpit's instrument panel, or the rules and procedures from manuals." Flight training had to combine these factors and accurately replicate physical sensations, mechanical signals, and official procedures. In Ed Link's "virtual flier," simulation and reality were conjoined. Observers often contend that the Link Trainer was a forerunner of virtual reality. Yet, as scholar Frank Cardullo said, "virtual reality is an oxymoron. Geometrically, a virtual image isn't located where it appears to be, while a real image is where it is. The more precise technical term would be 'virtual environment.'" With its first training sessions, the Link Trainer created an unprecedented form of a virtual environment and inaugurated a distinctive conception of flight.

The experiences provided by the Link Trainer may have been simulated, but their economic significance was real. "It was a low-hanging fruit," said Roger Connor, a pilot, and historian at the Smithsonian National Air and Space Museum. The merits of the Link Trainer in successfully unifying technological and psychological training for pilots became evident, and more profoundly so in establishing a new protocol for training in the aerospace sector. Most powerfully, the trainer consistently reinforced the flying fundamentals: position, direction, distance, and time.

Previously, pilots applied a step-by-step "1-2-3 method" to maintain straight and level flight. They adjusted the rudder pedals, control stick, and airspeed to influence how the aircraft's nose and wings moved. Even with this rudimentary method, pilots had to go against their instincts. The Link Trainer reoriented pilots' default dependence on their intuition. It trained them not to monitor their instruments but to trust them. Critically, the Link Trainer didn't merely mimic an aircraft's movement but prepared pilots to adjust instruments in response to varying conditions. The trainer added signals, subtracted distractions, multiplied viewpoints, and divided tasks. "It wasn't so much that the Link Trainer was an ideal instruction platform; it certainly wasn't," Connor said. "It was about knowing where you are now to visualize where you want to go next." The blue box was as much a skill-set trainer as a mindset trainer.

IN 1930, THE SAME year Ed Link opened his flight school, the Link Piano and Organ Company went out of business. His father, Edwin Link Sr., moved on to opportunities in real estate. For Ed Link, much depended on the newly formed Link Aviation. Around this time, the shy, wiry Ed Link met an adventurous journalist, Marion Clayton, who was assigned to profile him for the local newspaper. A romance took off, and Marion Clayton repeated throughout her life that she married her best subject.

Ed Link ran a government-approved airplane-repair shop in an unheated hangar to support his growing family. Between repair jobs, he built trainers by hand and ran his flight-training courses. He faced criticism that his trainer was mere "hangar flying" and could never simulate or supplant actual flight. But Ed Link's rebuttal highlighted his trainer's ability to cut cost, risk, and training time while producing proficient pilots. Case in point: When he trained his brother George on the Link Trainer in 1928, this would have taken 10 hours in actual flight, but the simulator cut training in half. George could safely solo after 4 hours of trainer time and 40-odd minutes airborne.

As the Great Depression ground on, Ed Link's revenues came mainly from air shows. Crowds applauded his skydiving, speed tricks, and balloon-bursting in the Pitcairn autogiro, also known as the "flying windmill." Ed Link tried to use his aircraft exhibitions to promote the trainer. In one instance, he and George towed the blue box on a borrowed Ford Model T to an event in St. Louis. The organizers refused to make it an exhibit, calling it a carnival ride. Coin-operated entertainment, once the province of the Link family's nickelodeon pianos, now included the Link Trainer.

Some circuses purchased blue boxes as amusement attractions. A barker beseeched the patrons in a state fair: "Just a quarter and you pilot a plane yourself." Marion kept the books and sold hamburgers at these shows. During the week, they lived off the leftovers. "Marion made more money selling hotdogs than I did flying airplanes," Ed Link recalled. An ever more desperate Ed Link was tempted to offer half his business to anyone who gave $500. He eventually sold 50 trainers to amusement firms, miniature golf courses, and only 2 to aviation firms, at $450 each. Even these purchases were for showroom demos rather than the actual purpose of training.

Frantic to bolster their finances, in 1933, Ed Link began experimenting with aerial advertising. He designed a forced-air mechanism using a borrowed monoplane to write words in the sky from under the aircraft wings. "I put a venturi tube in the slipstream of the airplane, and that created a vacuum," Ed Link recounted. "It was a player piano in the air!" The lights turned on and off from the ad text punched in piano rolls. He also equipped an aircraft with organ pipes. The flying billboards now had music. "Spaulding's Cakes Are Fresher," "Drink Utica Club Beer," and "Enna Jettick Shoes Are the Most Comfortable." The words beamed out of the night skies from 2,500 feet with an aerial roar.

As Ed Link once again proved his inventiveness over the skies of Binghamton, a political fiasco was ready to explode in Washington that would make the blind flight a national imperative. It started with a suicide club.

IN EARLY 1925, the US Congress passed the Air Mail Act, which required the postal service to privatize mail delivery. In response, bidders pooled their resources and formed the first aviation start-

ups. This burgeoning group of contractors took over the operations from the postal pilots in 1927. Some commercial carriers
called themselves "airlines" but were far from it; they didn't even
care about installing seats on their planes. A historian noted that
carrying airmail and freight—which paid three dollars per pound
for a 1,000-mile trip—was far more cost-effective than carrying
passengers. If a 150-pound passenger were the equivalent weight of
airmail, they would have had to pay "a prohibitive $450 per ticket"
(or around $8,000 in today's dollars). But cost differentials weren't
the only counterintuitive aspect of early aviation start-ups.

In 1928, a letter sent from Boston to Seattle would be routed by
one airline to New York, then flown to Chicago by another carrier, then to San Francisco by a different company, and finally to
Seattle. Mail navigated a "crazy quilt of routes," notes scholar Erik
Conway, resulting in sunk cost, lost time, and wasteful effort from
pilots. In 1929, the engineer-turned-US-president Herbert Hoover
dismantled this process through his postmaster general. Hoover
sought rational efficiency in mail delivery, and "shotgun weddings"
of consolidation among the many airline start-ups involved in the
mail business soon followed. More logical and lucrative routes
caught on.

No matter who delivered the mail, whether the Post Office or
the private companies, forced landings in lousy weather were an
enduring problem. Over 6,500 forced landings occurred in the nine
years the Post Office operated the airmail service. Their competition, after all, were the all-weather railroads. Some years earlier,
at a winter air meet in California, pilots wet their fingers in their
mouths. And they held them up to see if one side cooled quicker
than the other; call it a quick and dirty wind vane. When pilots
launched in poor visibility or bad weather, their lack of instrument
training made routine flights extremely dangerous. Conway high-
lights two types of weather-related accidents during the airmail
years. The first type of accident occurred when the pilots flew too
low in order to keep sight of the ground, often hitting trees, smoke-

stacks, and buildings. Contact flying in poor visibility was lethal. The second type of accident occurred when pilots stayed hazard-ously high in the clouds. Planes entering the clouds were prone to spin, nose-dive, and crash on the ground.

A year before his celebrated Paris trip, Lindbergh worked as a mail contractor and wrecked his plane twice in dense fog during routine runs. Despite the known dangers of poor-weather flying, Lindbergh's employers still expected him to deliver the gasoline-drenched mail promptly. Even with the 1933 completion of an illu-minated network of skyways and beacons, wicked weather buffeted profits and took lives. Conway observes a crucial point: if flight safety is paramount, one should stay on the ground. But safer fly-ing didn't mean *not* flying: "Aircraft that sit on the ground produce losses, not profits."

Concerned about airline profiteering, in February 1934, President Roosevelt canceled the private-airline mail contracts and directed the Army Air Corps to distribute the mail. Previously, in 1918, the army had temporarily delivered mail at a much smaller scale. This time, about 250 army lieutenants delivered mail in Boeing pursuit planes and Cur-tiss bombers. Before their dispatch to mail delivery, the young pilots had no reason to fly in poor visibility, and few had nighttime flying experience. Because aircraft training remained expensive and lethal, the Army Air Corps sent out a troop of gravely unprepared pilots.

Within the first month of the army's postal service, airplanes spun to the ground in dense rain and snow. In 70-odd days, 60-odd crashes took over a dozen lives. Roosevelt was under fire, and World War I fighter ace Eddie Rickenbacker declared it "legal-ized murder." An aerial demigod by then, Lindbergh denounced the president's actions as "unwarranted and contrary to American principles." Billy Mitchell, a US Air Corps legend, questioned the true proficiency of the pilots. "If any Army aviator can't fly a mail route in any weather, what would we do in a war?" Soon, the Army Air Corps halted its mail operations. It resumed delivering mail—but only during the daytime and in clear weather.

Two days after Roosevelt canceled flight contracts, but before the
Army Air Corps mail service started, Ed Link landed in Newark,
New Jersey, in dense fog. His associate Casey Jones had arranged
a demonstration of the Link Trainer for the Army Air Corps officers. The army ordered six trainers, at $3,400 each, which quickly
bolstered Ed Link's business. With the delivery of the Link Trainers to the army in June 1934, Ed Link kick-started a revolution in
aerial training.

In 1935, with government approval, Ed Link sold 10 trainers
to Japan. The following year, the 31-year-old Ed Link sailed to
Japan with Marion to provide instructions. After reaching Tokyo,
he found that the technicians had disassembled one of his trainers, apparently intending to copy its entire design. He refused to
help reassemble the machine, citing a lack of essential tools, and
returned home. In 1936, more trainer orders bubbled up from the
United Kingdom with the condition that the boxes be built in the
Commonwealth. The Soviet Union, Italy, Spain, China, and eventually over 30 more countries followed suit.

Now adopted by military powers across the globe, Ed Link's
engineering soon found its pivotal place in the lead-up to World
War II. With enough faith and persistence, what had once been a
circus thriller was finally a proper aviation device.

THE ROMANTIC POET Samuel Taylor Coleridge wrote that "shadows of imagination" required a "willing suspension of disbelief,
which constitutes poetic faith." Blind flight is a matter of blind faith
in instruments instead of rhyme. The Link Trainer was engineered to
encourage pilots into a willing suspension of disbelief; it regimented
real-time editing of sensory impressions gathered in flight. Pilots
interpreted sensory stimuli, even when they understood that they
were simulations. One aviator described the experience: "You start
with the box that frames the problem and into the box you put everything that contributes to trainee involvement—the controls, indica-

tors, etc. You omit from the box all that is not fundamental to the training situation. At least you do so to the extent that absence will not have a detrimental effect upon the trainee. Outside the box, you develop the instructor control . . . and then you connect the two areas through an appropriate interface. Finally, you insert the student into the box and close the lid." The key was to shut off external distractions so the trainees could be absorbed in the virtual environment.

Some resisted learning to fly in a box. "They didn't like the idea," Ed Link said. "If they just made them think it was an airplane, they enjoyed it better. So it was purely for psychology. It was the psychology of somebody getting in something that looked [more] like an airplane than like a box." Once "airborne" in the trainer, the pilots often thought the gauges were erratic and provided false information on their turns. "The turns did happen, of course," notes Conway; "that was the point." This new cognitive approach required pilots to "substitute their own psychological perceptions for the 'truths' created by their instruments," observes historian Timothy Schultz. But to better understand how the Link Trainer succeeded in rescuing pilots from depending on intuitions, let's look at a moment that initiated the instrument-flight era some years before Ed Link's fog landing in Newark.

VISION IS PERHAPS THE most reliable of our senses. Unlike with touch and hearing, without sight, our grasp of balance and direction is muddled. If you walk blindfolded, your sense of balance goes awry. Without sight to orient us, our inner ear can send mixed messages about where we are in space, resulting in a conflict of sensations. The pilots of the pre-instrument era flew dizzily in clouds. Lacking a visible horizon, they couldn't disregard their senses; they became disoriented and often spun out of control. Weather and vertigo were the archenemies of aviation.

As we saw earlier, William Ocker and Carl Crane's experiments in the 1920s provided initial insights into what happens to balance when lacking visual reference. But subsequent blind-landing experiments themselves lacked vision. In one approach, aviators lined airfields with tethered balloons to guide landing; another, led by the British and French air forces, laid an electric cable around the landing field for signaling. This attempt also failed, lacking precision transmitters and receivers to communicate with the aircraft. Other navigators attached long tail skids and weights to the planes that scraped the runway before touchdown to give pilots a physical signal. Eventually, a breakthrough emerged. In 1926, the mining magnate Daniel Guggenheim and his son Harry established a philanthropic fund to promote aeronautics. Their goals were clear: to ensure safe and reliable flights in all weather. They sought to identify ways to fly through the fog and address the general problem of blind flying. It involved a short, stocky, bald test pilot with a crooked nose.

At age 13, Jimmy Doolittle was adventurous. He built a wood-framed glider from the instructions in a 1910 issue of *Popular Mechanics*. Doolittle tied the contraption to a friend's car and successfully launched himself into the air, gliding briefly before he crashed. Doolittle also built an engine-powered glider with the extra cash made from boxing matches. It also flew successfully before a thunderstorm took it down. In 1925, at age 28, Doolittle earned the nation's first doctorate in aeronautical engineering from the Massachusetts Institute of Technology for studying how wind velocity affected flying characteristics. Later that year, he bagged another first: the Schneider Trophy, awarded annually to the winner of a race for seaplanes and flying boats. Doolittle topped the speed record in a Curtiss float biplane at 232 miles per hour. The next day he outraced himself at 245 miles per hour, more than quadrupling Glenn Curtiss's record from a dozen years earlier. Then followed an existential crisis. At the time of his win, Doolittle was still a first lieutenant in the Army Air Corps. With two kids and elder-care duties, becoming a captain was a distant dream. What next? Where was he going? Should he quit flying? Doolittle's phone rang. Harry Guggenheim borrowed him from the army to join their "fog flying" program.

The Guggenheim fund focused on five linked problems: first, dissipating the fog; second, effectively locating landing fields from the air; third, developing instruments to accurately measure how high the planes were; fourth, improving tools for flying in the fog; and finally, penetrating fog by light rays. The fund encouraged concepts to disperse the fog. Ideas included heating the atmosphere to above dew point; penetrating the fog with moisture-absorbing charged particles; spraying a drying agent over the fog to aid condensation; and, more mechanically, breaking up the clouds by propellers.

Doolittle and colleagues focused on improving the aircraft panels. They upgraded a Consolidated NY-2 biplane with an array of instruments. On September 24, 1929, Doolittle flew into the thick morning fog swirling over Mitchel Field in Long Island. Technicians pushed the biplane out of the hangar and prepared the radio system and landing localizers. Doolittle roared the 220-horsepower radial-piston engine. He taxied and took off into the fog, reaching 500 feet. He landed 10 minutes later, relying entirely on instruments.

A jubilant Harry Guggenheim encouraged Doolittle to repeat the feat, but this time with the cockpit covered by canvas sheet to achieve a fully blind flight. For safety, copilot Benjamin Kelsey sat in the open front cockpit with a full view. Doolittle lined up the aircraft for the liftoff from under the hood in the rear cockpit. The plane climbed a thousand feet and leveled. Flying some more miles east, Doolittle turned back to Mitchel Field, steadily descending and eventually landing. Kelsey sat still and relaxed, with his hands behind his head. "However, despite all my previous practice, the approach and landing were sloppy. The whole flight lasted only 15 minutes," Doolittle wrote. "This was just ten months and three weeks from the first test flight of the NY-2."

Doolittle's blind landing ended the seat-of-the-pants flying era. The Guggenheims shut down their fund's operation that year, declaring their mission successful. The age of instrument flight initiated by the Guggenheims then evolved from a stunt to a survival strategy, thanks to Ed Link. The Link Trainer provided calculated torture to those trained in it, teaching them to fight their instincts and trust their instruments. Doolittle and Ed Link's innovations worked in tandem: Doolittle's composite flight panel, outfitted with the latest instruments, enabled safe blind landings, and Ed Link's flight trainer, engineered to recreate all the sensations and technology of a cockpit, successfully trained a composite flier. But it took another deadly mishap—a political sandstorm—to elevate Ed Link's engineering to standard practice.

ON MAY 6, 1935, shortly after 9:00 p.m., Bronson Murray Cutting, a 46-year-old Republican senator from New Mexico, boarded a red-eye to Washington to debate a veterans' bill. Cutting flew aboard the Transcontinental and Western Air Flight 6—also called the Sky Chief service—featuring the twin-motored Douglas DC-2 and the roomiest cabin among passenger planes. The 14-seater had an 80-foot wingspan and cruised at 170 miles per hour. But some called the Sky Chief "too big to fly." Among the 10 passengers were a young woman with an infant daughter and a six-member Hollywood crew. When fueling was complete, the 28-year-old pilot and his 24-year-old copilot had no checklist duties. This practice wouldn't arrive until later that year, starting with the Boeing bomber B-17 Flying Fortress. As the wheels went up, Cutting went to sleep.

The route was routine. The weather, clear; the flight, smooth. But after crossing Wichita, a bad-weather alert squawked through the captain's staticky radio. Headwinds slowed the plane. The pilots prepared to fly on instruments, with a dense-fog warning issued for Kansas City. As the conditions worsened, Kansas City became zero-zero: no ceiling, no visibility. As the descent began, a TWA dispatcher directed the pilot to a government landing field in Kirksville, Missouri, 120 miles east of Kansas City. The conditions in Kirksville were a 1,200-foot ceiling and four-mile visibility. The flight heading toward the new destination was fog-shrouded, with misty conditions in even the few clear spots. The pilots attempted contact flying, with plans to track a concrete freeway. They dropped to 100-odd feet above the ground. At 160 miles per hour and 15 miles from the destination, the plane's left wing slammed into the top of a tree. The pilots lost control, and the aircraft rolled. At around 3:30 a.m., the Sky Chief crashed in flames on the edge of a pasture. Senator Cutting and four others died instantly.

As the congressional inquiries into air safety launched, so did the finger-pointing. Some briskly blamed the inaccurate weather reports, and others cited defective radio apparatus on the plane. Another party blamed the pilots; they were ill-qualified to fly, the hissing went. Even the beacons, located north of the crash area, were condemned. The Department of Commerce had lowered the beacon wattage to cut costs following the budget pressures of the New Deal. This cost-saving measure resulted in the "mushover effect"; the pilots couldn't pick up the low-intensity beam through the fog.

Royal Copeland, a Republican mayor in Michigan and later a Democratic senator from New York, led a sweeping congressional investigation. In May 1935, the Copeland committee held six days of hearings, eventually assembling 900 pages of testimony from the Post Office, the Weather Bureau, the Bureau of Air Commerce, the army, the navy, the airport-construction authorities, and the airline pilots' association. Within a month of the accident, the investigation concluded that the probable cause was the aircraft's "unintentional collision with the ground" due to poor visibility. Other contributing factors included the Weather Bureau's failure to predict hazardous weather, the aviation authorities' choice to grant takeoff clearance in Albuquerque, the pilot's judgment errors, and the Kansas City TWA ground crew's delayed decision to redirect the plane. TWA was penalized, and efforts commenced to structure and standardize the country's commercial air regulations.

The Copeland probe precipitated the 1938 Civil Aeronautics Act signed by Roosevelt. The act centralized aviation activities and diverted the Department of Commerce's responsibilities into a new, independent Civil Aeronautics Authority, which quickly divided itself further into two new groups. The first group, the Civil Aeronautics Administration, oversaw air-traffic control, safety programs, and airway development. The second, the Civil Aeronautics Board, centered on rulemaking, including pilot training and certifications. Although it unfortunately took a tragedy to compel government intervention, these new regulations for commercial air travel ushered in a new, safer age of flight. Meanwhile, as technological advances forged a new kind of cockpit, the pilot's role rapidly changed from throttle jockey to analyst.

WHEN ED LINK TOOK his first flying lesson in the 1920s, his instruments—the stick, pedals, dials, switches, compass, and rudder controls—were tools from aviation's earliest days. Soon, he'd learn to use the artificial horizon and the altitude indicator. In World War I, pilots benefited from the addition of the airspeed indicator. By World War II, the number of instruments grew, featuring a magnetic compass and devices for airspeed and turn-and-bank indication. By the 1950s, the cockpit panel became sophisticated enough to prevent stalls and crashes, crisscrossing a maze of pneumatic pipes and electrical controls. In the military arena, by the 1960s, a fighter plane alone had over 40 instruments. An aircraft's instrument panel could be adapted and configured for high-altitude fighting, reconnaissance, or ground-strike missions.

With the "instrument explosion" during this era of aviation, pilots took on an additional role: consulting analysts monitoring the systems under various conditions and constraints. "The day of the throttle jockey is past," noted a former head of the FAA, pointing out that the new breed of pilot was "becoming a true professional, a manager of complex weapons systems." In the more provocative phrasing of a retired air force colonel: "A modern pilot is like the corpse at a funeral: his presence is traditionally expected, but he doesn't do much." Pilots "rarely 'hand-fly' the aircraft as they comply with company directives to surrender control to the autopilot except for takeoff and landing. Sometimes even the landings are machine-flown." Test pilots complained that the barrage of information mentally drained them.

To realistically reflect the instrument explosion pilots would face in real cockpits, Ed Link's flight trainer, once equipped with rudimentary indicators, added directional and radio gauges. By the early 1940s, the trainer could simulate wind conditions with an analog computer. One 1945 news article declared that the trainer flew more like an airplane than an airplane. "Don't make the mistake of thinking this 'dummy' plane is easy to manage," aviator Assen Jordanoff wrote. "In fact, the trainer is far less stable and much more sensitive to control than a regular plane, for the simple reason that you can get much better training in a hypersensitive plane than you can in a slower and more stable machine."

Partly because the trainers could accurately reproduce the newest cockpit instruments, Ed Link's company tripled in size within a decade. Near the end of World War II, Link Trainer production hit its zenith: one blue box left the Binghamton assembly line every 45 minutes. While engineering transformed the pilot's role, societal progress and a pioneering vanguard of female and Black aviators redefined *who* could become a pilot.

ANY SYMBOL OF PROGRESS conceals its regressions, and aviation was no exception. Through the mid-20th century, women's opportunities in aviation reflected the roles women typically played in society. They started out as cooks and clerks, and recruiters told women that the aviation apparatus was similar to kitchen appliances. Women were hired for crucial support roles that would release men for military duties that women weren't permitted to perform. In public perception, women diminished their femininity by taking on the aviation roles of men. Jacqueline Cochran, who later broke the sound barrier, wouldn't exit the plane she piloted without touching up her makeup. That prompted a comment: "Though she flies like a man she hasn't become trouser-minded."

In the early 1920s, Amelia Earhart worked odd extra jobs as a stenographer and truck driver to fund her pilot's license. Her first airplane ride was 10 minutes long and cost her $10. Her lessons from aviator Mary Snook, the first woman to own a commercial airfield, cost her $500 for 12 hours. Earhart encouraged younger women to take up flying, even though this pursuit was unaffordable in an era before the Link Trainer. She argued, "They are simply thoroughly normal girls and women who happen to have taken up flying rather than golf, swimming, or steeplechasing."

In June 1928, a year after Lindbergh's flight, Earhart became the first woman to cross the Atlantic, as a passenger in the Fokker Trimotor *Friendship* piloted by Wilmer Stultz and Louis Gordon, from Newfoundland to South Wales. A year later, Earhart finished third after Louise Thaden took the top honors in the first transcontinental air contest for women; it was pejoratively called the Powder Puff Derby. Then, in May 1932, Earhart soloed across the Atlantic in her Lockheed Vega from Newfoundland to Northern Ireland, with winds stopping her from repeating the Lindbergh feat of five years before. Later that summer, she flew "her lovely red Vega" nonstop from Los Angeles to Newark in 19 hours, setting the record as the first woman to make this cross-country flight.

In 1930, United Airlines hired Ellen Church, a nurse trained as a pilot, as the first flight attendant. At the time, airlines wouldn't hire female pilots, and women were primarily employed in aircraft sales. While men could defray the costs of expensive flight lessons by working on airfields, women had to pay cash up front. Even for licensing, women were subject to a stricter standard than men; they needed to take an exam in addition to the requisite flight hours. Despite the additional barriers to entry that they had to navigate, female pilots were often the subject of tasteless jokes and crude comments about their competence. In one distressing instance, as historian Joseph Corn points out, a federal investigator rejected the possibility of engine malfunction in an accident and attributed the failure to neglect by the woman pilot.

For much of the 1930s and the 1940s, women's wartime participation transformed their role in the nation. Over six million women entered the workforce during World War II, taking advantage of remarkable opportunities previously unavailable to them, including as Link Trainer instructors. Their letters conveyed patriotism and passion, as well as newfound independence. "You are now the husband of a career woman—just call me your little Ship Yard Babe!" wrote one Indiana woman, adding that opening a checking account for the first time was a "grand and glorious feeling." A 19-year-old working at Douglas Aircraft Company as a blueprint manager wrote to her Marine Corps fiancé: "Imagine, [me], little Betty, the youngest in her department with seventeen people older than her . . . under her. . . . I am the big fish in my own little pond—and I love it." Another woman, from Cleveland, Ohio, told her husband in a letter: "Sweetie, I want to make sure I make myself clear about how I've changed. I want you to know now that you are not married to a girl that's interested solely in a home—I shall definitely have to work all my life. I get emotional satisfaction out of working and I don't doubt that many a night you will cook the supper while I'm at a meeting. Also dearest—I shall never wash and iron—there are laundries for that! Do you think you'll be able to bear living with me?"

In the summer of 1942, the US Navy formed the Women Accepted for Voluntary Emergency Services, a reserves program. The agency picked 74 women from the first class of WAVES to serve as Link Trainer instructors. In 1943, Ed Link himself was present to congratulate the WAVES graduates. Over the next three years, WAVES recruited 84,000 young women, many of whom were assigned to naval aviators training. The white, middle-class women were better educated than most of the seamen. In late 1944, the program admitted Black women only after pressure from activist organizations and approval from Franklin Roosevelt, just a month before his reelection to an unprecedented fourth term of presidency. The WAVES provided Link instrument training to novice naval gunners and aviators, including future US president George H. W. Bush, and recurrent training to experienced pilots.

The quasi-military Women Airforce Service Pilots program, or WASP, was formed in August 1943 from the merger of the separately organized Women's Auxiliary Ferrying Squadron and the Women's Flying Training Detachment. The WASP program selected 1,830 healthy 21-to-35-year-old women with a pilot's license from over 25,000 applications. About 1,100 women ultimately fought and 38 died on duty. Marion Stegeman Hodgson served as a WASP and was among the first women to become military pilots. She wrote to her mother: "The gods must envy me! This is just too, *too* good to be true. . . . I'm far too happy. . . . Honestly, Mother, you haven't lived until you get way *up* there—all alone—just you and that big, beautiful plane humming under your control." The WASPs flew every aircraft and received instrument-flight training, with some serving as Link Trainer instructors. In total, over 400,000 brave women from the various programs served the army, the navy, the marines, the Coast Guard, the Army Nurse Corps, and the American Red Cross.

Before and during World War II, the Link Trainer was also critical in expanding training for Black pilots. The US military had existed as a racially segregated institution since its inception. But Eugene Jacques Bullard dreamed of piloting machines in the air just as young white people did. Born in 1895 to a formerly enslaved family, Bullard ran away from his home in Georgia. In 1912, he boarded a freighter to Scotland, where he boxed and later joined a Black entertainment troupe in London. After a trip for a welterweight prizefight in 1913, he settled in Paris, joining the French Foreign Legion, toiling to receive his wings in 1917. As America entered World War I that year, Bullard tried to join the US Air Service and received a rapid rejection. But Bullard then proved his value in the French combat flight assignments he performed until 1919. After espionage entanglements during World War II and postwar stints as a security guard and a civil rights activist in New York, in 1959 the French government made him a Chevalier of the Legion of Honor.

Another pioneering aviator, Bessie Coleman, persevered in the face of racist and sexist US policy. In 1915, 23-year-old Coleman was working in a Chicago barbershop when a newsreel inspired her to take up aviation. All the US flying schools rejected the young woman because of her race and gender. So, she faked her birth date in her passport and set sail to Paris, determined that she could do anything French women could and better. After completing a 10-month course in 7 months, she received a pilot's license in 1921. Racial barriers proved insurmountable even as she excelled in aerial spins and swivels. Coleman's dreams to establish the first Black flying school in the United States crashed, and she pivoted to make her living off air shows. She refused to perform if Blacks and whites couldn't enter the grounds through the same entrance to the field. Her popularity won her enduring fame as "Queen Bess." Tragically, her life was cut short when her plane flipped in midair and threw her to her death in 1926.

Given the barriers to joining the military during this period, many Black aviators like Queen Bess made their living and honed their craft through air shows. One famous daredevil of the era was the Trinidad-born Hubert Fauntleroy Julian, nicknamed the "Black Eagle." In 1923, he parachuted into Harlem and received a ticket for causing a traffic jam. Julian bought a war-surplus seaplane and christened it *Ethiopia* after the kingdom he'd later defend. Unfortunately, his plans for a transatlantic solo flight failed due to his lack of financial backing.

In the 1930s, pilot William Jenifer Powell from Kentucky was considering not only how to join the field of aviation for himself but also how to open up the industry for all Black Americans. Powell enlisted in the segregated armed services during World War I. A toxic gas exposure in Europe ended his military career. Like Bessie Coleman, Powell then faced rejection from flying schools. With his engineering background, he knew aviation was an instrument of world commerce. Powell envisioned Black people taking roles across the industry. With two business associates, he founded the Bessie Coleman Aero Club. Its members included James Banning, who in 1932 was the first Black pilot to fly coast to coast. Banning's mechanic, Thomas Allen, fixed a rickety plane with surplus parts. Reacting to the old Curtiss motor, Banning said, "While originally it developed 100 horse-power, I have reason to believe that some of the horses are dead." Banning and Allen called themselves "Flying Hoboes," completing their trip from Los Angeles to Long Island in 41 hours and 27 minutes over three weeks. Landing "with the gracefulness of a bird, rolling only a few feet before stopping. I had made it!" wrote Banning, "and, immediately, presto-change, a great transformation occurred. My head erect, eyes to the front, shoulders squared, I was a different man."

After decades of pioneering feats by Black pilots, in 1938, the US military developed its civilian pilot-training program across colleges and extended training to Black cadets at historically Black colleges and universities. Pilot Charles Alfred Anderson was hired to head the Howard University program in Washington, DC. The following year, at an isolated campus in Tuskegee, Alabama, founded initially by Booker T. Washington, the US military created a new training program with Anderson as the lead instructor. In January 1941, the War Department started the all-Black 99th Pursuit Squadron, which was to be trained at Tuskegee. Later in May, Eleanor Roosevelt visited the campus. She asked to take a flight with one of the pilots, saying, "I always heard that colored people couldn't fly airplanes." Anderson flew the First Lady on a Piper Cub even as Secret Service remained uneasy about the ride. This highly publicized visit ensured the White House's support for Black pilots, sending the nation a strong message about their proficiencies and patriotism.

Some war-propaganda posters now featured Black pilots, in one instance with the message "Keep us flying! Buy War Bonds." Mrs.

Roosevelt maintained long-term correspondence with some Tuskegee pilots. In a summer 1942 letter, one airman wrote to the First Lady, alluding to the Link Trainer: "Your letters and gifts have been very inspiring and have prompted me to try to be a better soldier. My work is very interesting. . . . Soon this short radio course will be over and I'll be of some service to Uncle Sam." "Chief" Anderson directly instructed hundreds of cadets at the Tuskegee Institute. And over time, just under 1,000 Black cadets trained on Link Trainers. In 1943 and 1944, the 99th conducted multiple attacks on enemy aircraft, resulting in some of the most successful days of the European campaign. By the end of World War II, Tuskegee Airmen earned decorations for their magnificent record of nearly 1,600 combat missions.

While these developments were underway, a group of army pilots was scaling heaven in Binghamton, thanks to Ed Link's newest and stellar creation.

IN 1939, ED LINK began developing a Celestial Navigation Trainer to help aircrews simulate a bombing mission and improve their accuracy in night raids. The critical element of celestial navigation was a device called a sextant. Ed Link developed a training sextant by modifying an instrument used for maritime navigation and applying insights from his close friend (and decorated naval officer) Philip Van Horn Weems. The sextant displayed the angular distance between the stars and the horizon without electricity or the need for stillness. The finely numbered metal arc formed one-sixth of a circle, which gave the device its name. A minute, one-sixtieth of a degree, measured a nautical mile.

Although he came from the conservative naval tradition, Weems was an iconoclast with big ideas about applying maritime navigation principles to flight. Weems met Lindbergh and introduced him to the maritime sextant for the 1927 flight to Paris. Lindbergh plotted his Great Circle course with piloting, guided by fixed ground

reference points during daylight. And at night, he used dead reckoning to deduce his present location based on already flown distance and course. Weems thought Lindbergh could have traveled
at higher altitudes, reaching the destination earlier using less fuel.

In his influential book *Air Navigation*, Weems wrote that understanding the fundamentals of navigation could have saved countless lives in the early days of aviation. For instance, in the Dole
Derby of 1927, the winner, *Woolaroc*, used celestial and radio navigation to reach Honolulu. *Aloha*, the second-prize winner, used a
marine sextant. Those who perished in the *Golden Eagle* flew past
the northern part of the Hawaiian Islands. They lacked navigation
skills and eventually ran out of fuel on the breathtaking blue. A
decade later, the 39-year-old Amelia Earhart and her navigator,
Fred Noonan, disappeared in their silver Lockheed Electra over the
South Pacific. For their pioneering 29,000-mile journey, the pair
traveled without navigation or Morse-code communications. Eventually, they lost visibility, fuel, and their lives. Weems devised an
almanac convenient for celestial sightings, which proved helpful for
Ed Link's next invention.

Ed Link once equated his trainer's benefit to the value a globe had
in presenting geography. But the Celestial Navigation Trainer was
an artificial paradise. The 45-foot-high silo looked like an abandoned granary, but its chief feature was a star-studded synthetic
sky, like a planetarium. A chicken-wired cupola projected over 400
electric stars overhead, including the principal constellations of the
Northern Hemisphere. Motor and gears precisely rotated the pre-
tend celestial sphere 360 degrees in 24 hours.

An expanded Link Trainer at the center of the silo accommo-
dated the bomber crew. An instructor on the ground floor set the
flight conditions and plotted the crew's performance using the
tracker on a chart. The navigators supplied the flight course using
the view of the constellations and radio aids. They pinpointed their
"plane's" location against the moving stars produced by a collimated effect. The Celestial Navigation Trainer projected 250,000

square miles of photographed terrains as a mosaic on a silkscreen, varying altitudes from 3,500 feet to 35,000 feet. The British Air Force used board models with an elaborate jumble of hand-painted color prints in England. These celestial "fixes" oriented the trainees, who set the targets on a checkerboard of images the trainer "flew" over. Once the trigger was pulled, a glow on the screen indicated whether the mission was accomplished or not.

Over time, the Link Trainer and its descendants enabled the lifesaving technological and psychological advancement of learning to fly without leaving the ground. Nonetheless, questions about the training procedure remained: What behaviors must be trained, and to what criteria? What is the optimal mix of theory, practice, and reflection? How does one evaluate the effectiveness of training and make the best use of costly assets? Building on these considerations, systems engineer Valerie Gawron offers generic training criteria that she has developed over the years, calling them the "ten C's": completeness, clarity, conciseness, consistency, compactness, communications, competence, correctness, constructing from previous knowledge, and staying current. Arranging these Cs becomes critical for trainees to effectively transfer what they have learned in the trainer to a real scenario.

Advanced versions of the Link Trainer mimicked artificial storms, fog, icing conditions, rough air, glare, and the zero-zero condition that led to Senator Cutting's death and many other unfortunate deaths of passengers and pilots. The trainers became sophisticated with so-called upset-recovery training, in which pilots reacted and adapted under extreme circumstances. And with newborn computing power, flight trainers were on their way to becoming flight simulators.

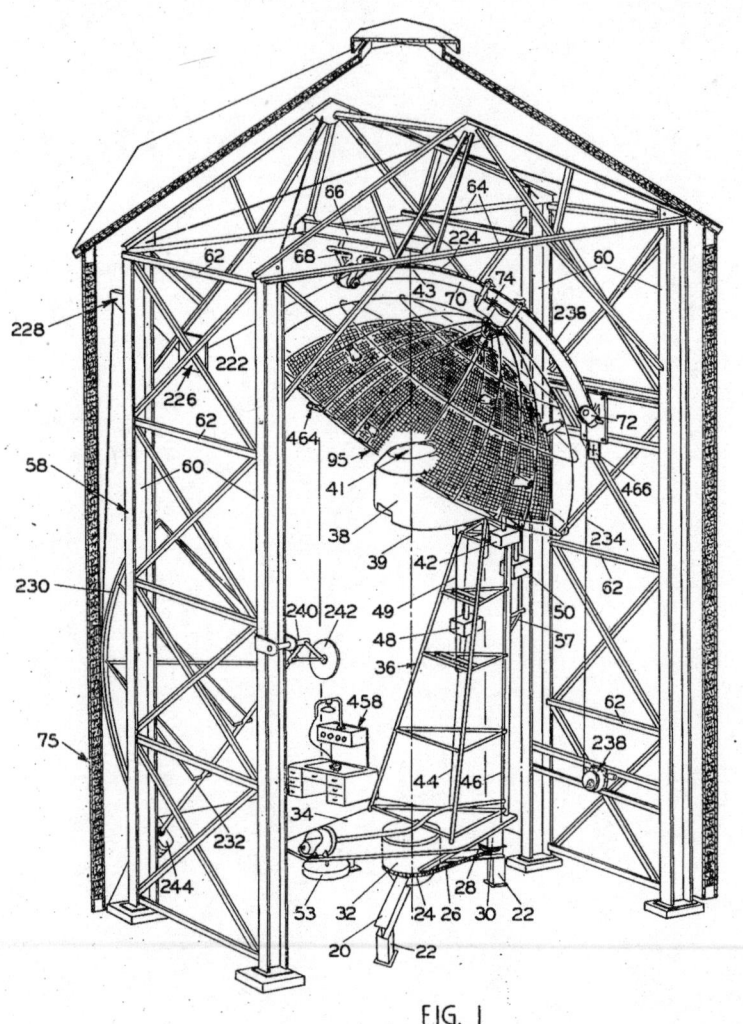

FIG. 1

EDWIN A. LINK
INVENTOR.

BY Donald T. Hillier

ATTORNEY.

Chapter 3

Yaw

TAKEOFFS ARE VOLUNTARY, AND LANDINGS ARE MAN-
datory. Pilots James Morton and Maurice Watson were well
trained in this maxim.

At half past midnight on March 30, 1967, Morton, a trainee captain, guided Delta Flight 9877 out of the gate at New Orleans airport. His instructor, Watson, accompanied him as captain of this
routine proficiency check. Their airplane, the Douglas DC-8 aircraft, had completed a regular passenger trip from Chicago earlier in the day. Three Delta employees and an inspector from the
Federal Aviation Administration were also aboard this short practice ride. The training agenda was straightforward: takeoff, circle,
and touch down with a simulated two-engine out on one side of
the aircraft.

Nearly equal in training, Morton had completed over 17,000
flying hours, and his instructor, Watson, was approaching 20,000
flight hours. The cockpit atmosphere was relaxed as the flight
began to taxi at 12:40 a.m. At 12:43 a.m., the aircraft reached its
critical engine speed. When it was too late to reject the takeoff,
Morton and Watson simulated the first engine failure by powering
it down. A minute later, they simulated the second engine failure
during the climb. At 12:46 a.m., with both "failed" engines idling at
1,200 feet and 200 knots, Watson informed Morton that the rudder
power was lost. With a warning light pulsing on the dash, Morton

quickly dropped the nose of the plane to steady it, leveling at 900 feet. Morton reduced the speed to 180 knots and flaps on the wings to 25 degrees, too much and too early. The aircraft lurched to 1,100 feet. Captain Watson reminded Morton of what to do.

12:48 a.m.

Watson: Don't let that thing get below a hundred and sixty [knots].

Watson: Ball in the middle, Jim.

Whatever it takes, put 'er in there now.

Morton: Get my landing gear for me.

Under the jet-black skies and a nearly full moon, New Orleans sparkled. Morton quickly turned the aircraft toward the illuminated runway two and a half miles away. Fellow crew members encouraged Morton, "Okay, Bud, looks good," "How 'bout that," and "Now we're straightened out." Watson took over the landing checklist and lowered the landing flaps.

12:49 a.m.

Watson: Wing flaps, landing flaps . . .

Morton: Call my airspeed for me.

Watson: One forty.

[Sound of engines beginning slight spool up]

Watson: One thirty-five.

Watson: See, you're letting her get. . . . 'ut the rudder in there. . . . You're getting your speed down now; you're not going to be able to get it.

Morton: Uh uh.

They had five-mile visibility; the landing plan was a 2.5-degree glide path, a regular touchdown. But the airplane's angle was 3 degrees, and Morton didn't compensate for the increased drag from the flaps by adding speed for the descent. Watson didn't cor-

rect the trainee captain. Morton decreased the descent by raising
the nose angle instead of increasing the power. He lost control.

12:50 a.m.

Morton: CAN'T HOLD IT BUD.

Watson: Naw, DON'T, let it up, let it up . . . let it up!

[End of recording]

The aircraft hit a large tree about 2,300 feet short of the runway,
then slashed through two more. As it hovered out of control, the
plane sliced through houses, damaged vehicles, and pulled down
power lines. The D-8 aircraft scattered its tail and engine parts,
sparking fires and gouging a 30-foot crater. Next, it slammed a
railway embankment with a metallic screech and skidded toward
a Hilton hotel, where it exploded, destroying about 50 rooms. The
sea of fuel and the wreckage killed 9 high school girls, one of whom
was blown out of the building. The girls were among 23 students
who had financed their trip to New Orleans from bake sales, car
washes, and babysitting. The total death toll was 19, including
those on the plane.

The accident report concluded: "It is obvious from the total evi-
dence that the causal area lies in the human element." It identified
the probable cause as improper supervision by Captain Watson and
Morton and Watson's inappropriate use of flight and power controls. The report reasoned the possibility that "because of the near
equal status of the two pilots, the instructor was more hesitant to
take control of the aircraft," which he should have done earlier in
the flight. The report also noted that Watson's intense workload
and minimal rest likely contributed to his inappropriate supervision. These compounded factors resulted in an avoidable and "non-
survivable accident." The fact that a simulated proficiency test
could have spared 19 lives didn't go unnoticed. Forced to confront
their complicity in this tragedy, Delta and other airlines turned to
flight trainers as the exclusive training method. But airline compa-

nies could have reached this conclusion three years before Watson and Morton's mortal mistake.

On another New Orleans–based flight in February 1964, Eastern Air Lines Flight 304 took off toward Washington Dulles. The crew aboard the Douglas DC-8—the same aircraft flown by Watson and Morton—lost control over Lake Pontchartrain. Unable to yoke up the plane from a full nose-down position, the aircraft plunged into the lake. All 58 passengers died, including a recently wed 21-year-old stewardess who had successfully fought the airline's policy of not employing married flight attendants. After the accident, investigators recovered the flight recorder from under 20 feet of Lake Pontchartrain silt. The unit was too damaged to yield a cockpit transcript. The investigators studied the plane's maintenance records and other functioning and failed DC-8s for the cause. They concluded that the accident resulted from a faulty pitch trim compensator destabilizing the aircraft. In a cruel irony, the part that caused the crash was due for service the next day. Notorious for their mysterious and menacing part failures, the Douglas DC-8s earned the monicker "death cruisers." But mechanical errors weren't the only contributing factor. Investigators also concluded that the pilots' lack of emergency preparedness contributed to this preventable disaster.

As aviation tragedies go, the 1964 and 1967 New Orleans crashes were darkly telegenic. These dramatic and seemingly avoidable disasters sparked public outcry and influenced the FAA to change its flight-training requirements. Since then, commercial pilots have trained exclusively on simulators to practice emergency protocols.

IN THE EARLY TO mid-20th century, all aviation accidents seemed like wicked problems. Most safety-related issues were "designated mysteries"; a single part failure could trigger a consequential chain of events. Glitchy engineering and human error are a wicked combination, and it seemed impossible—even politically risky—to assign blame for accidents after the fact.

For most of aviation history, flying made enormous demands on a pilot's cognition. In the late 1930s, one physician reported a condition called "aeroneurosis," a nervous disorder in which the stress of flying, mainly over oceans, fueled fatigue. In the years and decades that followed, airlines introduced policies to curb pilots' exhaustion, including backup pilots and crew replacements. Engineering enhancements also helped alleviate the demands on a pilot's attention. The jetliners of the 1950s boasted additional safety features and automated meters that made them more manageable than the older piston-engine aircraft. Even with these advancements, a pilot's work remained exhausting and complex, with life-and-death consequences.

Since their development in the early 1940s, flight-data recorders have been considered controversial. The Air Line Pilots Association derided the "black box" recorders (which are, in fact, bright orange) as "nothing but a mechanical spy." Indeed, the black box intends to record, if not spy on, a pilot's actions in the cockpit. Engineered by David Warren, the son of an Anglican missionary who was killed in an unresolved flight crash, the black box could track a plane's motions. The data allowed investigators to reconstruct what happened in the event of an accident. With essential work by engineer James "Crash" Ryan, the American flight recorders were developed by Lockheed Aircraft Services and, oddly, Waste King, the garbage-disposal company, and General Mills, celebrated for cereal brands. These early models recorded and transmitted the plane's speed, altitude, vertical acceleration, and compass data. The black boxes came in a heat-resistant, crash-protective, battery-powered case.

The recording units were popularized in the 1950s after the "Comet crashes," in which three new De Havilland planes were all catastrophically damaged in the air. Soon after, the recorders became an unassailable feature of the aircraft. In the debates preceding their deployment, developers decided on the best place for the equipment: the aircraft's tail, usually the last part damaged in a crash. As one professional said, they "never saw an airplane back into a mountain."

The black box is sometimes called the "indestructible machine." As scholar Greg Siegel points out, the black box was protected "for every conceivable calamity, its mission was to go to hell and back." The institutionalization of black boxes, Siegel notes, coincided with Cold War surveillance, interception, and decryption. In 1966, a year before the Morton and Watson accident, the FAA regulated minimum performance requirements for the black box. They determined that an aircraft's flight recorder should withstand impact shocks up to 1,000 g's for five milliseconds. For perspective, at zero g's, we feel weightless; at 1 g, we can firmly stand on the ground due to the Earth's gravitational field. But at 100 g's, a human body is crushed. The FAA also mandated that recorders endure 5,000 pounds of force and an impact penetration of 500 pounds dropped from 10 feet. Black boxes had to withstand corrosive fluids such as oil and fuel for 24 hours and be able to stay under seawater for 30 days.

The flight recorders weren't that different in concept from Ed Link's motorized recorder on the instructor's desk to track the fictional course of flight training from no place to nowhere. As the black boxes were untangling the mechanical mysteries of plane crashes, the blue boxes unraveled another aspect: psychological training for disaster preparedness.

IN THEATER LEGEND Sanford Meisner's technique, an actor strives to "live truthfully under imaginary circumstances." Within the Link Trainer, pilots were asked to achieve a similar goal. In the 1950s, analog processors sped up simulators' capacity for aerodynamic calculations. Increased computing power enabled an elaborate Link training system for an all-weather fighter aircraft. It involved "8,300 engineering drawings (using 180,000 square feet of various kinds of paper), 8,150 pounds of structural steel, two and one-half tons of aluminum, 5,000 pounds of electrical cable, enough power to heat, comfortably, two 6-room houses, enough air-conditioning to cool two 6-room houses, enough electronic tubes to build 82 television sets or 282 radios, 250,000 feet (about 44 miles) of wiring (equivalent to the amount required for 125 well-lighted homes), 352 transformers of various types, 10,636 resistors, 93 servo motor generator sets, 85 simulated aircraft instruments and 225 parts from the actual F-89 aircraft." Once completed, the simulator was 10 feet high, 24 feet long, and 23 feet wide. The simulator accurately reproduced the experience of locating and destroying adversaries with 1,000-pound bombs at the peak speed of 530 miles per hour. Within the Link Trainer, fighter pilots learned to live truthfully under imaginary circumstances. They gained skills that might one day prove invaluable.

While the devices of the 1940s and 1950s emphasized *flight training*, the digital computers of the 1960s and onward gave rise to the more popular and durable term "flight simulation." While all squares are rectangles, not all rectangles are squares. Similarly, all trainers are simulators, but not all simulators are trainers. The Variable Anamorphic Motion Picture, developed by Link engineers in the 1960s, used distortion lenses and mirrors to display widescreens from 70 mm color projectors. The VAMP showed videos of actual takeoffs and landings filmed at the Chicago O'Hare airport and later programmed them into pilot exercises. The films were so convincing that the movie adaptation of the best-selling 1968 novel *Airport* by Arthur Haley used Link's footage, echoing the Link Trainer's origins in entertainment during the Great Depression. In 1977, the first six-dimensional motion system was launched— capable of pitch, roll, yaw, surge, sway, and heave—a hexapod with hydraulic jacks and limbs. The "synergistic six" swiveled to mimic, however imperfectly, the scenes and sensations of actual flight.

In the 1990s and 2000s, "simulation syndrome" set in with the development of a slew of specialized simulators. Soon, the tail gunners, radar operators, and command posts each had their own simulators. These advanced systems were equipped with precise aerodynamic effects and facsimiles of the instruments that trained operators in different, specialized scenarios. These digital simulators did not merely train pilots how to fly. Instead, they taught pilots complex techniques, including managing an aircraft's systems or exercising aerial combat maneuvers, such as dogfighting. The machines produced repeatable and reliable actions in the pilots by faithfully reproducing human sensations and machine controls in code.

Today's simulations have transcended Ed Link's idea of using only the necessary ingredients to produce the essential experience. The general goal—even an FAA requirement—of modern simulators is to have as many elements identical to the actual flight conditions as possible. Ideally, trainees should feel no difference between flying a real aircraft and "flying" a simulator. The more convincingly features are reproduced, the better the training transfers from the simulator to the cockpit. Or, as the scholar Sherry Turkle has written, "In the culture of simulation, if it works for you, it has all the reality it needs."

THE GENERATIONS OF PILOTS trained at FIT Aviation, Florida Tech's division for flight instruction, have offered living proof that simulated training produces real-life success. Originally built in the 1940s to train navy and marine aviators, FIT Aviation still occupies part of the beachside airfield southeast of Orlando, once a flat cow pasture surrounded by a salt marsh. Nearby West NASA Boulevard teems with aerospace and defense contractors' buildings. On a periwinkle February day, professor Donna Wilt waited at the Center for Aeronautics and Innovation entrance, ready to take me on a flightless flight. "My dad was an engineer in maintenance and trained on the blue box. So I just had that knack," Wilt said, walking to a hall of simulators.

"This is an infinite runway," she said of a visual emanating from a wraparound projector from a specialized simulator. It was a thrill ride from the nearby Disney World. "Normally, you only have a few seconds to land because the runway runs out. But this one goes on forever." This simulator allows beginners to hone the essential motor skills for hand, eye, and foot coordination through repetitive practice. Another exercise involves landing with different crosswinds. Trainee pilots can compare their results, like video gamers vying for the highest scores. Wilt described those landing contests as "hangar parties" where the simulators were used for entertainment. "It's fun to have a party with a bunch of flight instructors."

A century ago, Ed Link focused on delivering an affordable training platform. These days, simulators offer what Wilt described as "over-the-counter coaching." Many products strictly compete on price now. "So basically, we get a Microsoft Flight Simulator with a really cool box around it," Wilt said, referring to the software series older than the company's most famous product, Windows. "Even the instruments are not real. They are simulated." The trick with modern flight simulators is determining their mission requirements and how their fidelity is tailored to the syllabus. "If you are training for knob switching, this is perfect. If you are teaching them procedures and checklists, it's ideal. And basic instrument skills? You can't beat it." Wilt described another simulator, noting how procedure trainers and full mission trainers for the same vehicle can coexist in the same course and place. As Ed Link understood, a simulator's realism affects a student's ability to adapt the training to fly. That's the idea behind the "transfer of training." A positive transfer means the students can effectively carry the knowledge, skills, and abilities from the virtual environment to the actual flight. A negative transfer occurs when habits, emotions, and prior experiences inhibit that crossover.

Even the low-cost, mass-market simulators that Wilt demonstrated rendered a blade of grass in fine detail. "That's a high fidelity, sure, but it's also a gimmick for the look and feel and to increase sales," Wilt said. The question of real proficiency gain for students remains an open question among flight instructors. Putting someone in a swanky simulator doesn't mean that person will be a competent pilot. How real should a simulator be? And how much detail is enough? There are many forms of fidelity. *Physical fidelity* is about how a trainer looks, feels, and sounds. *Cognitive fidelity* relates to the trainer's reproduction of the physical and psychological workload in piloting. And *motivational fidelity* reflects whether the device is friendly and engaging in a way that encourages students to adhere to training.

A vivid visual experience is engaging, but it may not necessarily teach pilots the best techniques for handling, for example. That kind of know-how comes from practicing the yoke or rudder controls to achieve a natural effect, such as a coordinated flight turn. Despite the effectiveness of simulators, some practitioners believe that no simulation is complete without the experience gained on a real plane. In recent years, the FAA has increasingly valued simulator training over actual flight experience, which remains an expensive and risky alternative. The crosswind landing trainer Wilt demonstrated at FIT Aviation was purely for personal practice, not FAA training credit. However valuable for process training, it was an altogether stationary experience, lacking motion cues that Ed Link thought were indispensable for flight trainers.

In fact, the debate over whether motion cues are necessary for flight training has been a long-standing one. During the post–World War II slump and eagerness to sell more units while saving costs, the flight-training industry hatched a plan to do away with motion, the defining feature of the Link Trainer. Newer builders argued that motion trainers didn't provide g-forces as experienced during flight. Ed Link disagreed. In 1947, he wrote that the "total lack of vertigo" in stationary trainers was a detriment. Ed Link believed that "because vertigo is definitely experienced in the air and part of one's training is to overcome this sensation," if these physical cues weren't present, instrument training would not be satisfactory.

Not every engineer agreed with Ed Link's convictions about the necessity of motion. Five years later, the B-47 jet bomber simulator replicated the six-engined Boeing Stratojet. But the simulator dubbed the "sitting duck" was stationary and had no motion or acceleration effects. One company executive said, "The pilots become so preoccupied with the realism of their flight duties in the simulator that they don't appear to miss the sensations of the motion." But Ed Link never changed his mind about the inferiority of motionless simulators, and the US military seemed to agree. In the 1960s and 1970s, the military was reluctant to accept simulators that didn't result in the "pucker factor," slang connoting the stress, specifically rectal contraction, affecting pilots in full alert and fear. Motion fidelity is linked to the adrenaline response in pilots—palpably, a whole-body response that could affect their ability to fly. One Australian wing commander stressed: "If there's no pucker factor, it's no bloody good, and it doesn't fly like the real aeroplane."

The motion-no-motion debate occupied center stage during the 2010 merger of United Airlines and Continental Airlines. Continental trained its crew on fixed-base simulators, requiring no motion. In contrast, United regarded full-motion training as highly realistic and necessary to prepare its crew for aircraft stalls, upsets, engine malfunctions, and landing problems. Continental made an economic argument to the FAA, noting that no-motion systems were cost-effective in developing crew proficiency. In contrast, United strongly opposed that idea, as Continental tried to expand the practice to United. Mergers, just like marriages, are as much about dealing with incompatibility as they are about cherishing compatibility.

At FIT Aviation, I was ready for my first ground lesson on an FAA-approved advanced aviation training device. Ron Mock, a retiree from the air force turned training and simulation manager, was setting up the instructor console. It was a single-engine, four-seater utility plane, Piper Archer, ready to take off from the Melbourne airport.

"The master switch is on the lower left over there," Mock said. "Yep, that one."

I turned on a switch for the engine. "All right! Who wants to fly?" I said.

"Wait, the engine stopped running," Mock said, pausing. "The key must be off."

"I thought we were already in the air," I said.

"Not yet," Wilt said. "You turn the left knob there."

(Relief.) Everything now seemed under control as I steered the aircraft for liftoff. It was going to be a smooth flight, I told myself.

"Now . . . what's that ticking noise?" Wilt asked.

"What ticking noise?" I asked.

(Silence.)

"It's the hydraulics," Mock answered.

"Right," I mumbled. I pulled the stick to lift the nose. As I gained altitude, we heard a screaky beep. "What did I do?"

"I have no idea. That's not a sound you hear in a Piper," Wilt said. "Oh . . . so I see the problem. This says we're about to crash, but we obviously aren't."

(Confidence boost.) The sound was gone a moment later.

"Off we go into the wild blue yonder, climbing high into the sun."

The top view of the terrain was clear—some scattered clouds. The world looked different from the flight deck. Roads, trees, houses, and shores. The scenery was generic but realistic. There was no pressure of flying a real aircraft, and no lives at stake, including mine.

"We don't know our angle of attack directly," Wilt said, "but things like airspeed are critical for estimating it. I'm watching that. With airspeed and the power setting, I have information about the plane's drag and thrust, which will determine our performance and descent rate," Wilt added, checking off all the readings. "That looks good, and that one's good too."

"I feel like I'm on a blind flight," I said. "I don't know what's happening, but it looks like something's happening."

"You're banking a little, so level," Wilt said. "Look out that wing and see how much ground is above. See where the horizon is? Now, look at this side. And if they're the same, then you're level."

"Yeah. I think I am."

We settled at 3,000 feet as the plane whirred.

(A few minutes later.) "Let's lose some altitude," she said. "See where the horizon is above the nose? Let's level this."

I brought it down to 2,300 feet.

"I see the runway," I said, adjusting myself like a Top Gun correcting for a carrier landing.

"No. Don't land there," Wilt instantly corrected me. "That's the shuttle landing strip. Kennedy Space Center."

I thought it was an excellent time to pull in a barrel roll like a barnstormer, but just then, Wilt said, "Oh, it's raining."

"I inserted a weather effect," Mock said with a grin.

I giddily steered the stick, all set for a thunderous roar on the runway.

The touchdown was anything but that.

I landed the Piper on a small puddle.

"That was a nice touch."

AS THE FLIGHT-SIMULATOR INDUSTRY has grown, so have the tendencies to "gamify" the experience. As early as 1944, Ed Link received a suggestion for a pure game based on the blue box and the Celestial Navigation Trainer. "Not that I wish to throw any cold water, but I have had some experience in using technical subjects and gadgets to amuse people, and found that the public is very fickle and sometimes unreceptive to amusements that might even improve their own good," Ed Link wrote in a letter. "In other words, it is a fickle and unpredictable business which we, as technically minded people, get enthusiastic over, and sometimes misjudge human nature."

Today's flight instructors can attest to the perhaps uneasy union of entertainment and technicality in simulators. Donna Wilt recalled a recent experience trying out a new full-motion simulator. "I flew it right into the Grand Canyon," she said. "The whole time I was down there, it was going 'Terrain. Pull up! Terrain. Pull up!' whereas I was having fun." Then she offered the controls to her copilot, a retired airline pilot. " 'Let's get the hell out of there,' he said. He didn't like it at all." Whether a mega-franchise movie or a cheap trainer with off-the-shelf simulator software, on-screen effectiveness is a function of realism. Being highly realistic doesn't necessarily mean the simulator is effective.

Visual effects are "euphemistically called 'hybrid' or 'empirical' techniques, and more candidly called 'grotesque hacks,' " Alex Seiden, a technical director for seven Oscar-winning films, has written. "The dinosaurs of *Jurassic Park* were not 'simulated' any more than pre-World War II Los Angeles was 'simulated' for *Chinatown*." Since the mid-1990s, when Seiden wrote this, gaming has dramatically influenced Hollywood and aviation, with people accustomed to much richer, more realistic imagery.

But how does one balance purpose and presentation? This is a perennial question. Seiden discussed an aspect of this problem related to designing dinosaurs for the screen. "Often, a precise simulation would not only be more complicated but would also be aesthetically undesirable; for example, the scale of dinosaurs in *Jurassic Park* changes dramatically from shot to shot and sequence to sequence," as he put it. Seiden's point applies to modeling and simulating flight terrains. Mountains cannot resemble pyramids, and the greenery must be blended with gross imperfections to convince the eye. Such visuals should combine with the pilot's senses to provide a stable reference and understanding of the operating environment.

Like Seiden, former Link simulation engineer David Gdovin understood, perhaps better than anyone, the balance between purpose and aesthetics when designing simulated environments. In 1995, Gdovin and fellow engineer David Peters cofounded Diamond Visionics to produce "virtual worlds in real-time." "Every single building in this view is procedurally generated from data," Gdovin said, explaining their latest software. Their simulator displayed the city of Seattle with 14,000-foot Mount Rainier meditating in the background. Their software engine ingests data and satellite imagery and runs them in real time to produce the simulated vision of Seattle. "The program gives us the shape of the footprint. It gives us the location of the footprint. It gives us the orientation of the footprint. It gives us the height of that footprint," Gdovin said. "And bingo! We take all that data and render Seattle 60 times a second from the raw source data we're given." This means that, true to life, the Diamond Visionics simulator never displays exactly the same terrain twice. Graphical programming is generative—a tight dance of polygons and pixels, allowing three-dimensional scenes to be built and modified on the fly, now a common feature in many applications.

Indeed, some of Diamond Visionics' earliest work happened in parallel with simulation for laparoscopy, a minimally invasive surgery in which a thin telescope tube with a camera is inserted to see inside the body without needing a large incision. The simulation to train surgeons in laparoscopy required special effects of lighting, wetness, and reflection from the internal organs, as well as realistic images of blood, tissues, and surgical instruments, precisely what a surgeon would see. Gdovin's team adapted the idea for military simulations: replacing the surgeon with a warfighter, changing blood to soil terrains, and all the rest.

"So what's our destination, San Francisco or Honolulu?" Gdovin asked Jason Petrick, one of his lab engineers.

"Let's go to Honolulu," Petrick said, and he launched a descent scene.

"Here's the out-the-window day scene," Petrick said. A few more mouse clicks. "Here's the night scene. And here's the infrared."

Looking at the demo was like riding a magic carpet over Oahu. The reflections of the sky on the ocean and waves appeared Pixar-quality.

"Do you notice what's happening to the runway here?" Petrick asked Gdovin.

"Put runway edge lights every 30 feet," Gdovin said, asking for a replay.

"It just takes a few moments to roll in that new data," Gdovin said. "Boom! We got edge-lit runways, and it's as simple as that."

My attention turned to the waves in the ocean.

"We're working on shadows of the waves, just to make it more realistic," Petrick said.

Jonathan Richards, a quality-assurance engineer who tests for glitches and bugs, was sitting across from Petrick. "Yeah, everything should be realistic," Richards said, zooming in and out of midtown Manhattan on his monitors. Richards scrutinized the rendered images for lag and hardware interruptions that could make the experience unpleasant. The result, "barfogenesis," is a motion sickness caused by visual media.

Back at Petrick's monitor, the descent scene in Honolulu was advancing. The distant flight strip came into view. As the aircraft banked, the vistas varied dazzlingly with the vantage point. With a slight turn, the reflective textures of the ocean changed. Buildings, billboards, and their shadows swept past, conveying the essence of what philosopher Paul Virilio dubbed a "dromoscopic vision," a vision born from speed, as if on a race course. "The world flown over is a world produced by speed."

Gdovin asked for some weather effects.

"How about some rain on this paradise?" Petrick asked.

Gdovin had a twinkle in his eye.

"How about a blizzard?"

The power and ability of Diamond Visionics' simulations are literally dizzying. "Think of any red brick building, and I can render that red brick building," Gdovin said. But the quality of the visuals isn't the only meaningful component of their simulations. "Now, the pilots don't care whether it's a red brick building. They only care about not hitting it," Gdovin said. "The stakes are very high in this kind of simulation. Lives depend on it."

To determine whether the produced visuals are meaningful, one should consider three focal levels of perception. The first level of perception is image-referenced and linked to the simple identification of things, say, a tree. The second level is world-referenced, which requires a contextual understanding of the visuals; for example, the tree is on a mountain. The third level is knowledge-referenced, involving the importance of the context; say, the tree on the hill is approaching closer and faster, which may mean the pilot's losing altitude. All this is to say, specifying a simulator's level of visual detail is as much an engineering challenge as it is a computing requirement. How, then, does one establish the minimum acceptable detail for a simulator?

The high-performance geospatial demos and immersive experiences available today may have been unthinkable to Ed Link and his contemporaries. But, the challenge of specifying a simulator's level of detail would have been quite familiar. In the early 1940s, he grumbled about the requirements set by the government for the trainers. Ed Link felt that the regulators had "been adding one thing after another which makes the Link Trainer considerably more costly and more complicated, which takes more equipment and men to produce, and considerably complicates the production problem." He saw no value in producing a complicated trainer for novice training and intended to simplify the instrument trainer further. "The cockpit will be square like a box, will have no wings or tail group. There will be complicated and hard to produce wind drift. There will be no useless gadgets such as a leveling device," he wrote. "I am cutting out everything that is not being used on primary instruction." While you can find highly complex demos like Gdovin's, many of the simulators that Donna Wilt demonstrated at Florida Tech were designed with Ed Link's original objective of simplification.

Flight-simulation professionals often argue that "it's not how much you have but how you use it." As scholar Eduardo Salas observes, this idea has created a standstill in aviation because advances in simulation and simulators have overshadowed the training itself. Learning develops through an array of influences, including how the learning environment is designed and how able the instructor is. Training is ultimately a cognitive and behavioral activity, and we "must abandon the notion that simulation equals training and the simplistic view that higher fidelity means better training."

WORLD WAR II CREATED business stability for Link Aviation Devices in the 1950s. But at his business's apex, Ed Link began losing interest in flight simulation. Ed Link complained that the ingenuity in aviation was not what it used to be in his earlier days. "I sort of lost interest in the space age," he said, "when they computerized everything, although it was a very natural and important step. But I was more interested in originating than looking at a computer."

In 1953, Ed Link stepped down as president and became chairman of his company's board. A year later, Ed and his brother George sold the company. In the years to come, Link's future holding companies, including the Singer Corporation, produced a striking array of simulators designed for everything from commercial flights to weapon systems. They included specialized tasks for everything from vertical takeoff and landing to cockpit procedures, radar landmass simulators, submarine trainers, power plant simulators, and industrial simulators, with far more advanced instructor consoles. Now there are generic simulators for a suite of tasks beyond aviation, including for cars, trucks, ships, locomotives, surgeries, and bomb defusions—with lifelike sounds and experience.

"I started as a grease monkey with dirt under my fingernails," Ed Link told a reporter, "and now after making my fortune as a swivel chair executive I am back as an experimenter."

Ed Link had moved on to a different medium.

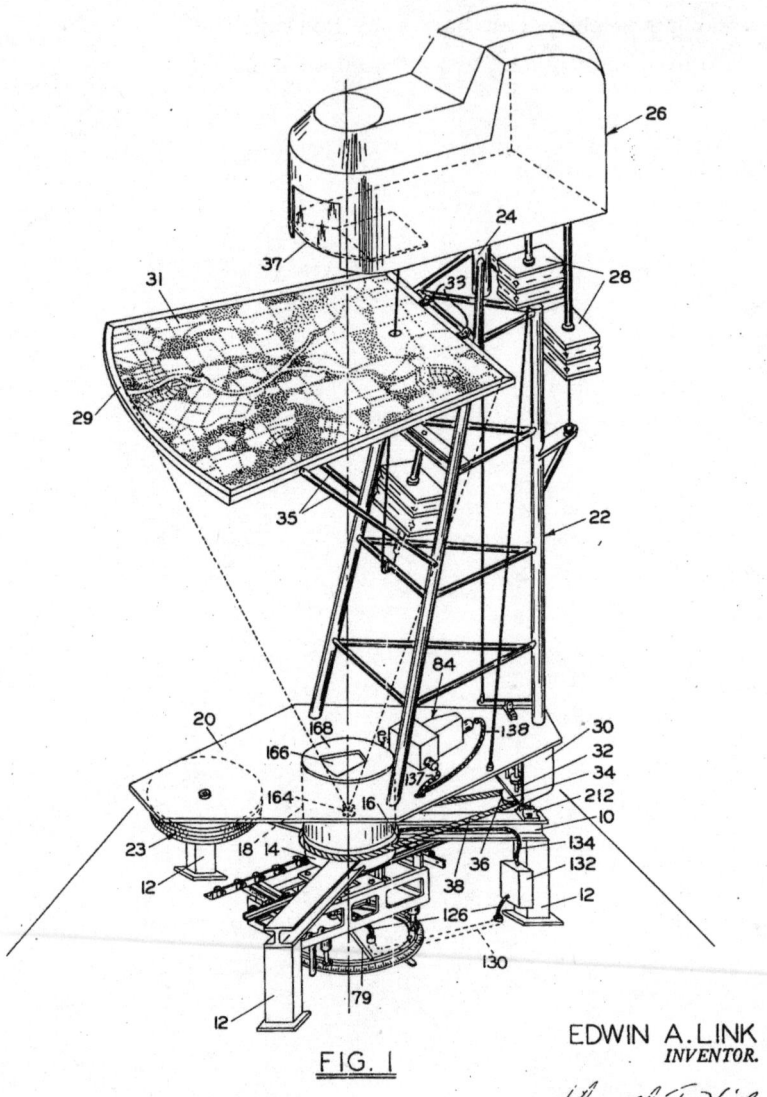

FIG. 1

EDWIN A. LINK
INVENTOR.

BY *Donald V. Hillier*
Philip S. Hopkins

ATTORNEYS.

Chapter 4

Surge

ON NOVEMBER 21, 1962, PRESIDENT JOHN KENNEDY WAS edgy. He called his science adviser Jerome Wiesner and NASA leaders James Webb, Robert Seamans, and Hugh Dryden into the White House Cabinet Room. He told Webb, the NASA chief, that the lunar mission should be NASA's prime focus. "Jim, I think it is the top priority. I think we ought to have that very clear."

Webb, a former attorney, expressed concerns about the unknowns and was pessimistic about the project's feasibility. As an argument brewed between Kennedy and Webb, Wiesner, an engineer, came down on Webb's side: "We don't know a damn thing about the surface of the moon, and we're making the wildest guesses about how we're going to land on the moon." Seamans, also an engineer, agreed.

Kennedy remained undeterred. The "Soviet Union has made this a test of the systems. So that's why we're doing it. So I think we've got to take the view that this is the key program," he said. "And the rest of it, God . . . there's a lot of things we want to find out about. Cancer and everything else." Webb tried to reason with the president that securing scientific leadership with the space mission would be a better bet than pure political symbolism in the simmering war of ideas between capitalist and communist countries.

"But you can't," Kennedy fired back, "by God, we've been telling everybody we're preeminent in space for five years, and nobody believes us."

"All right, sir, but let me say this," Webb replied. "If I go out and say that this is the number one priority and everything else must give way to it, I'm going to lose an important element of support for your program and for your administration."

"By whom? Who?" Kennedy asked.

"By a large number of people," Webb said.

"Who? Who?"

"Well, particularly the brainy people in industry and in the universities who are looking at a solid base," Webb said. "I'd like to have more time to talk about that because there's a wide public sentiment coming along in this country for preeminence in space."

"Yeah, but see, you got to prove you're preeminent," Kennedy retorted. "Unless this is the way to prove you're preeminent."

ELECTRIC TEAKETTLES CAN BE finicky, even for someone who helped put humans on the moon.

"Let's give it two more minutes, the water's not hot yet," Gerald Gene Abbey said, jiggling the switch up and down. He picked a bag of organic black tea from the tea chest and said: "Yeah, we don't want anything *in*organic."

Abbey was born in 1928. The son of a construction worker, he grew up in Flint, Michigan. "I was never much of a scholar," he said with a coy smile. "I had too much farm work to do." After serving in the Marine Corps, he studied engineering at Lawrence Tech. He failed college algebra but eventually got the hang of it. "It's easier the second time around," he said, adjusting his fatigued blue T-shirt, its divided pocket holding his reading glasses and a pen. After college, Abbey got a job in the General Motors truck and coach division. His first project was to develop fluorescent lights for a bus operator in New York City, the Fifth Avenue Coach Company. Then he was assigned to the multispeed-windshield-wiper project. "Gee, maybe we could make them go slower when it wasn't raining so hard," Abbey said.

In 1956, Abbey was recruited by Link Aviation, then under the ownership of General Precision Equipment. Abbey's first impression of Ed Link came from Tom Watson Jr., the second president of IBM. "Ed Link is a guy I really admire," Watson told Abbey. "If he wants to do something, he just does it." Abbey worked for the Link corporation for over 50 years under different ownerships. He started as a field engineer for the B-52 Stratofortress simulator. Then he led the prototyping of the simulator for the Lockheed Electra, a turboprop airliner from the late 1950s that is still used today.

In October 1957, the year after Abbey began his work with Link Aviation, the Soviet Union launched *Sputnik*. The 185-pound artificial satellite reached the Earth's orbit, earning the USSR a vital win in the space race and stirring American anxiety. Soon after, the newly created NASA launched Project Mercury, aimed to launch a crewed spacecraft into Earth's orbit and return it safely, ideally before the Soviet Union. After countless sessions on Link's Mercury simulator, on May 5, 1961, Alan Shepard piloted the first crewed Mercury flight aboard the *Freedom 7*. From 1961 to 1966, Project Gemini, a mission that extended spaceflight capabilities to be used in the Apollo program, relied on even more advanced space simulators. Abbey was the program manager for the Gemini simulators at Link. "We had evolved as an organization to take on challenges like that," Abbey said. "Newton showed us the physics, and Link did the engineering."

Kennedy's national goal of landing humans on the moon and returning them safely to Earth necessitated increasingly realistic simulations. For Link's Apollo simulator program, Abbey was the associate program manager for systems integration and testing. A key challenge was the complexity of managing the spacecraft configuration, which was compounded by the absence of vehicle-performance data. For aircraft simulation, one builds on test data; when needed, one could test out a real aircraft. With the Apollo simulator, data had to be predicted, and it had to be correct, for lives were at stake. In other words, the "flight" part of the flight test had to be entirely simulated.

A crucial hurdle for the simulation was computing power. In the late 1950s, the first programmable machine ENIAC—Electronic Numerical Integrator and Computer—was a grouping of over 1,500 relays, 6,000 toggle switches, 7,200 crystal diodes, 10,000 capacitors, 19,000 vacuum tubes, 70,000 resistors, and 5 million hand-soldered joints. It weighed over 30 tons and occupied 1,800 square feet. ENIAC consumed 150 kilowatts, "spurring rumors that every time it was turned on, lights dimmed in Philadelphia," one observer wrote. With the dawn of the digital computing era in the early 1960s, IBM machines with speeds reasonable for flight simulation cost around $2 million each. "This was an intolerable cost in view of simulator economics at that time," noted engineer John Hunt, who pioneered the development and use of digital computers at Link. Even if the visual presentation was a matter of basic arithmetic, rapid regeneration of those pictures to simulate motion and control loading cues required unobtainable speed and storage.

In the mid-1960s, Abbey and his team worked with 16-bit microprocessors that could process 25,000 instructions per second. Realistic simulations were the only way to prepare for the moon mission. Sending a flight crew to the moon in order to train for going to the moon wasn't an option. "We got faster through the course of the project, though," Abbey said. "Essentially, we evolved with the available tools." Link rented a spacious hangar at the Binghamton airport for the simulator assembly. The region's chronic cloud cover provided a natural setting for specific tests requiring subdued light. But when the northeast blackout of 1965 disrupted their work, the simulator settings had to be redone from scratch. As designers at NASA changed their minds about the contingencies of the mission, Abbey's team had to respond swiftly to the ever-changing details of the vehicle configuration. To address these alterations, a Link simulator draftsman reconciled spacecraft changes from NASA contractors to keep the reams of production-floor drawings current. "That was our version of rapid prototyping!" Abbey said.

Abbey's learning curve was steep as Link's computing software migrated from machine language to programming languages, changing the protocols for systems integration and testing. "It was just the daily grind of improving, testing, and installing the math models," Abbey said. Then, in January 1967, an electrical fire on the Cape Canaveral launchpad destroyed the command module, killing Gus Grissom, Ed White, and Roger Chaffee, the Apollo 1 astronauts. This tragedy cast further doubt on the already-delayed Apollo mission, and investigators questioned the readiness and safety of the project. As NASA made the necessary changes to the launchpad and spacecraft, the simulators had to be updated again.

Undaunted by the setbacks and the stress-packed environment, the Apollo mission brought together renegades whose work would define engineering and human history. Gene Abbey recalled a bus ride through the large campus of an Apollo contractor. An unassuming engineer sat next to him. The man was blond, crew cut, fit, and two years younger than Abbey. They spoke about airplanes and their hometowns. The civilian test pilot was from Wapakoneta, Ohio. In the early 1950s, he had flown 78 combat missions as a naval aviator in the Korean War, where Abbey had served as a marine. At the end of the ride, Abbey received a scribble on a piece of paper from that person.

His name was Neil Alden Armstrong.

HISTORIAN STANLEY GOLDSTEIN HAS cataloged the human impulse to transcend Earth. Ancient Greeks imagined a powerful water jet that could propel humans to the moon. One 17th-century idea involved a set of rockets propelling an aerial chariot; another, an iron wagon pulled toward the moon by supermagnets. Other dreamers envisioned filling two huge vases with smoke, sealing them airtight, strapping them under the arms, and allowing the smoke to propel one toward the moon. Writer Jules Verne imagined man's journey to the moon in his 1865 novel, *From the Earth to the Moon: A Direct Route in 97 Hours, 20 Minutes*: a columbiad space gun firing a passengered projectile into outer space. Although written a century before the Apollo mission, Verne's satirical vision of an international space race and a crewed rocket to the moon may be the most prophetic vision of space travel in its time.

Many individuals imagined embarking on the half-million-mile round trip to the moon. Still, for those tasked to realize it, training and teamwork were the two most essential skills. Both were paramount in building the mental and physical proficiency to respond to life-threatening situations. During Project Mercury, designed to determine whether humans could function in space, the astronauts spent over 700 hours—over half of their training time—in Link simulators. For Gemini, the project further exploring human capabilities in space and the long-term effects of space travel, astronauts spent 67 percent of the total training time in Link simulators. And in the Apollo program, the combined training time in simulators for command and lunar modules was about 25,000 hours. This was in addition to the dramatic sessions on Boeing Stratotanker, dubbed the "vomit comet." The cargo plane made 40 to 50 steep climbs and dives in a two-to-three-hour period. Each parabolic arc simulated weightlessness by floating the astronauts for 20 to 30 seconds. "The final simulation before a mission was much like a graduation ceremony, except that instead of going out into the world to get a job, we had the task of landing an American on the Moon," wrote engineer Gene Kranz, NASA's chief flight director during the Gemini and Apollo programs.

If Apollo's astronauts were fearless, their simulator had to be peerless. But the simulators hardly looked heroic. The oddly angular boxes housed a jumble of electronics, mechanics, and optics: nearly 700 switches, over 400 circuit breakers, 64 kilometers of wire, and 26 subsystems. Astronaut John Young nicknamed it the "great train wreck." While the original Link Trainers for airplanes did not simulate the external environment, the Apollo simulators required a sophisticated visual experience. Stellar visuals in the lunar module were created by a luminous ball, a miniature descendant of

Ed Link's Celestial Navigation Trainer. The two-and-a-half-foot spherical mirror made a specular reflection. With over a thousand illuminated ball bearings on its surface, each miniature version of an actual star created a perception of faraway celestial fields.

While debates about the merits of motion or motionless simulators remained in the world of commercial flight, for the Apollo moon landing, engineers determined that astronauts had to know what motions, sounds, and forces of acceleration to expect on their mission. Link's engineers achieved force loading on the ground using motors, springs, and sliding shafts to give the "real feel" of the rocket. Precise sound cues indicated a drop in cabin pressure or thrusters firing in countless conditions. Instructors could freeze, fast-forward (by 30 times), and fine-tune practice conditions for astronauts training in the simulators. Of course, these operations required immense computing power. It took four computers to run a simulation—three to drive the equations of motion, the physical variables described as a function of time, and one for the visual system. The host computers were 24-bit machines, each with 350 kilobytes of memory. The 40-ton, 30-foot-tall Link Apollo simulator was a fragile contraption, with 10 tons devoted to lenses, mirrors, and film alone. Engineers had to employ a turboprop airlifter, Douglas C-113 Cargomaster, and 14 semitrailers to transport the simulator.

As one historian aptly writes, "Apollo gave new meaning to the word 'astronomical' in terms of cost and accomplishment." But it also enabled a "world of precreated experience" where reality was "just like the sim." Ultimately, the steep cost and brainpower required to create the simulator were worth it. The Apollo 11 trio put in 14-hour days, practicing at least 50 landing scenarios over three months, making several round trips to the moon. These landing scenarios were akin to practice symphonies of systems that mix photography, dialogue, sound effects, and score into a coherent motion-picture track. All the training was a prelude to an extraordinary cinematic event that premiered on a balmy July day in 1969.

Neil Armstrong, Buzz Aldrin, and Michael Collins reached the moon after a 109-hour journey—12 hours longer than Jules Verne's uncanny estimate from 1865. When the *Eagle* landed, the moon's temperature was 200 degrees Fahrenheit. Armstrong descended from the lander to walk on the Sea of Tranquility. From Houston, astronaut Bruce McCandless, the mission-control capsule communicator, said, "Okay. Neil, we can see you coming down the ladder now." Armstrong adjusted his water-cooled space suit for a "pretty good little jump" onto the fine-grained lunar soil. He stepped off the lunar module, proclaiming an 11-word cosmic soundbite: "That's one small step for man, one giant leap for mankind." Fifty seconds later, while walking on ground previously untouched by humankind, Armstrong's words commended the Link engineers. "It's even perhaps easier than the simulations at one-sixth g that we performed in the various simulations on the ground," he said. "It's actually no trouble to walk around. . . . Okay, Buzz, we ready to bring down the camera?"

APOLLO 11'S SUCCESSES RESULTED from tenacious testing and rigorous refinements over many years. The late engineer and philosopher Walter Vincenti called this iterative practice "normal design." Engineers work to understand how the system operates, what considerations would form a desired solution, and the feasibility of accomplishing it. After years of education and practice, design work becomes intuitive. Engineers can comfortably probe "how the device works" and "what it looks like," a symbiosis of function and form. Through "normal" engineering, the airplane's design has stabilized. That is, cockpit is usually up front and tail in the rear. The design of an airplane has rarely changed, so it's easy to differentiate an airplane from a helicopter if you look at their operating principles: how they work and how they look. This understanding creates an accepted practice. If an engineer veers from those established operations, the result can be a "radical design." The design is more radical if the technology's working principle is fundamentally altered compared to how it looks. Such unique changes produce engineering know-how that may or may not depend on science.

An excellent example of radical design is a weight-loss project at NASA. In early 1965, the 14,850-kilogram lunar lander prompted worries that it was too heavy to do its job. NASA immediately ordered a design diet and formed the Weight Control Board. Grumman, the company that designed and built the lunar module, launched a war against the lander's excess weight. One approach, the "Scrape," searched for opportunities to lighten up the lander's parts. The second approach, the Super Weight Improvement Program, or "SWIP," involved a radical reenvisioning of the lander. A team of engineers second-guessed the entire design, which was 95 percent complete and involved contributions from 20,000 separate contractors. By September 1965, the Scrape project whittled away 45 kilograms from the structure. And by the end of 1965, Scrape and SWIP eliminated 1,100 kilograms of excess weight. A radical change involved using aluminum-mylar heat foil instead of bulky thermal shields. This choice demanded skillful machining and testing, leading to more weight-reducing changes and an oddly angled assembly.

The original Link Trainer was also a radical design from an unlikely source: player pianos. Instead of being born from an existing simulator design, the trainer's organizing force gave rise to new scientific insights into safety and human behavior. Engineering is often deemed derivative and subservient to pure science, even when new sciences are routinely developed from engineering. But the Link Trainer enabled new sciences and organizational ideas for learning, not vice versa. And within a generation, this radical device became an institutional fixture in aviation.

THE APOLLO 11 MISSION—ENABLED partly by the radical design innovations, coordinated efforts across teams, and ongoing training—was a crowning achievement for America during the Cold War era. While the Soviet Union and the United States remained locked in a battle of innovation, not all nations could replicate the engineering feats that launched these nations into scientific history. In 1967, a French politician commented on why Europe lagged in the space race. The deficiency was not so much in brainpower, his reasoning went; it was with "organization, education, and training." In other words, France lacked systems management.

Scholar Stephen Johnson has studied the impacts of systems management on aerospace missions, which invariably involve interlocking practices of planners, engineers, scientists, and bureaucrats. Well-organized processes can markedly improve the performance of launch vehicles, ballistic missiles, and spacecraft. In the decade between the 1950s and 1960s, Johnson points out, the success of such projects following a systems management approach almost doubled, with cost and schedule overruns reduced by 3- to 10-fold.

For adventurers, systems engineering procedures can seem dry and tiresome. But think of it as the tedium of lifesaving flight-safety checks and instructions that precede the thrill of flight. "We can lick gravity," as the sentiment goes, "but sometimes the paperwork is overwhelming." If the concept of flight is about six degrees of freedom, then processes that support it present six degrees of constraint. That's why the aerospace sector lives with a paradox. Innovation and bureaucracy breed each other. The safety-critical products and processes that produce exciting missions must be infallible. "Spacecraft that fail as they approach Mars cannot be repaired," Johnson writes. "Hundreds can lose their lives if an aircraft crashes."

Consider the acronym-heavy, six-step procedure applied to Apollo configuration management, which systems engineers designed to precisely track and control hardware and software modifications. The first step was a Preliminary Design Review (PDR). The second step was the Critical Design Review (CDR). The third step was the Flight Article Configuration Inspection (FACI). The fourth step was the Certification of Flight Worthiness (COFW). The fifth step was Design Certification Review (DCR).

And the sixth was the Flight Readiness Review (FRR). This routine might sound creatively deadening. But these steps were essential to realizing the mission, from parts to the performance and progress reports. People fascinated by the "moonshot" have focused only on the impossible destination and seldom discuss the procedures and discipline that undergirded the mission. No wonder many of today's so-called moonshots often end up as moonshine, only intoxicating advertisements rather than integrated action.

The aerospace sector has relied on systems engineering as "insurance for technical success," Johnson observes. These methods are invaluable when planners have an incomplete picture, with poor objectives, communication, and coordination. Sometimes a mission may combine all these deficits. "Military officers demanded rapid progress. Scientists desired novelty. Engineers wanted a dependable product. Managers sought predictable costs. Only through successful collaboration could these goals be attained," Johnson writes. Success with space rocketry necessitates using systems engineering tools, however mind numbing or time consuming, in the spirit of the Latin expression *per aspera ad astra*—through difficulties to stars.

ONE MAN WHO DEEPLY understood the indispensability of systems engineering—and the tragedy when it is ignored—was Frank Hughes, the former chief of NASA's spaceflight training division. Hughes joined NASA as an instructor in 1966. On the first day of his job at the Kennedy Space Center, the first box of the Link Apollo simulator was delivered. Hughes's career at NASA evolved with the Link simulators through the Mercury, Gemini, Apollo, Apollo-Soyuz, and space shuttle missions leading up to the International Space Station program.

"We were lucky all the time," Hughes told a reporter. "I used to joke that God spent about half of his time just with NASA on those early missions." The early stage of the Apollo program had "no training rhythm to it," Hughes recalled. "The crew walked in the door and set the stage of what they wanted to do. There was no such thing as a lesson plan or objectives or whatever. You trained until you flew, and it just filled up the time. If things delayed, then you trained more, and you just went on like that."

One of the most critical elements of the training was the ability to simulate potential problems that astronauts could encounter. During the Apollo program, the Link engineers updated the simulators with over 1,000 exercises for malfunction scenarios. "The fuel cells could fail, tanks could leak, all that sort of thing," Hughes remembered. Simulations could be sped up and slowed down with advanced visuals and many improved capabilities for every step of the mission rehearsal. The star fields were even more credible. And the simulators now contained the exact measurements of every spacecraft element rather than estimates.

The value of such almost identical features became apparent on Monday, April 13, 1970. Apollo 13, the third lunar landing mission, was 55 hours into its mission and 200,000 miles from Earth. The spacecraft was under the command of Jim Lovell, with John Swigert as the command module pilot and Fred Haise operating the lunar module, all trained as engineers. Meanwhile, after an intense and exciting day, Frank Hughes was driving home from Kennedy Space Center, already thinking about the training regimen for Apollo 14.

Suddenly an oxygen tank exploded in the service module, creating a sequence of breakdowns on Apollo 13. Lovell, Swigert, and Haise couldn't identify the problem. Soon after, the crew heard several more explosions. In an emergency call to mission control, Lovell uttered the most famous, inauspicious words in space history: "Houston, we've had a problem." Hughes raced back to mission control. He and the rest of the team quickly got to work on the simulator, continuing the effort for five restless nights. In the first three hours after the oxygen tank explosion, the crew lost over one-half of the oxygen reserves. Returning the trio safely to Earth was now *the* mission.

With the command and service module *Odyssey* crippled, *Aquarius*, the lunar module, became the "lifeboat" for the astronauts to operate for 90 hours, double the number of hours the module was designed to work in an emergency. As the spacecraft hurtled away from Earth at 2,000 miles per hour, preserving the module's power was paramount. New procedures had to be researched, recorded, and revised. The Link simulator operated nonstop as the ground crew and standby astronauts rapidly reworked their ideas. They tested scenarios based on the remaining onboard resources.

With oxygen reserves depleting, the astronauts' challenge was compounded by the fact that their simulator sessions didn't perfectly reflect their reality. Although high in fidelity, the simulations didn't have the cloud of particles or debris striking the spacecraft that the crew had to contend with onboard. And there were further problems that the simulator hadn't anticipated: maintaining uniform temperatures within the spacecraft. Apollo 13 needed passive thermal control. A tricky maneuver first attempted in the simulator, nicknamed the "barbecue mode," slowly spun the spacecraft lengthwise for even exposure to solar heat. With a disabled command and service-module computer and the lunar module lacking the essential software, the slow-motion roll had to be done manually. Lovell controlled the forward-backward movement, and Haise handled the left-right motion of the spacecraft, rotating the spaceship at three revolutions per hour. This technique saved the spacecraft from being heated on one side and frozen on the other. And finally, thanks to the ground simulator, the crew learned a method to abandon the disabled module, which enabled the astronauts to plummet safely through the Earth's atmosphere. As the astronauts commenced reentry, the procedures were anything but orthodox. With orange-and-white parachutes inflated over the Pacific, they splashed down near Samoa.

For NASA, Apollo 13 was a "successful failure," Hughes noted, and the Link-built simulators out of Binghamton proved that they were the "unsung heroes" of the rescue mission.

"THIS IS THE HALL of Ones and Zeroes," said Susan Sherwood, pointing me to an area in the old ice-cream factory, now the Center for Technology & Innovation, in Binghamton. Sherwood, the center's director, described the workings of vintage IBM computers and printers. "The very creators are now restoring them," she said. Local company uniforms and coffee mugs were also on display. On the other side of the room sat a pair of pianos, one of which was a Link player unit. The coded roll of a "Fats" Waller medley rang out the bubbly "Boogie Woogie" and "Canteen Bounce" followed by the cheery "Fuddy Duddy Watchmaker." The gears turned, the wheels spun, the paper rolled, and the keys mysteriously went up and down. "Makes you feel kind of creepy, don't it," wrote the engineer-turned-writer Kurt Vonnegut in his debut novel, *Player Piano*. "You can almost see a ghost sitting there playing his heart out." The old box had come to life.

The Center for Technology & Innovation is a theme park of antiquated technologies with a vast collection of devices and manuals from IBM, Link, and General Electric, all formerly local businesses in Binghamton. The center's all-volunteer program TechWorks! aims "to showcase regional technology in action—an effort that begins from the inside out." For Sherwood, "what's past is prologue." Along with the volunteers from TechWorks!, Sherwood restores the unsung heroes of the past—machines like the player pianos, IBM computers, and Link Trainers—and reveals the profound lessons these technologies might hold for our futures. "Shall we go see the flying piano now?" asked Sherwood.

Ed Link rigged his first flight trainer from a piano not far from TechWorks! on Water Street. Sherwood took me to the center's working antiquary, a hangar-looking building that doubles as an exhibit hall. The central idea of TechWorks! is operational preservation. Since 2006, Sherwood said, the center had returned three dilapidated Link Trainers to working status. Over two years, volunteers refurbished the 1942 blue-box trainer using new-old stock parts. Repairing the instructor station took another two years. The synthetic trainer returned to an operational state with refreshed servo motors and microcontrollers. Now visitors to TechWorks! can enjoy an "immersive" spin on the device that revolutionized aviation.

Several retired Link engineers have dedicated themselves to preserving these technological treasures in the working antiquary. John Lash recalled working for Link as a career-defining experience. "The company was one big happy family of people," he said, adjusting the belt on his cargo work pants, remembering his daily commute to Binghamton. "From the guy who swept the floors on the second shift up to the president who ran the company, they were all neat people to interact with," Lash reminisced. "Over the years, that changed; it became more businesslike. It all came down to the bottom line."

Lash quickly discussed an electronic component with Carl Mazzini, a fellow Link veteran conducting a circuit test. "Even people living in Binghamton don't fully realize how significant their town is," Lash said. "There are stories behind everything here if you had people to tell them." And Mazzini said, "I saw an old washing machine over there. The company that became Whirlpool started here," referring to the vintage cast-iron Cataract from the early 1900s. "That's something I didn't know."

Lash lamented a long-vanished coupling between a company, its community, and the country. That connection, he argued, inspires meaningful work. "It's one of those things where I think ignorance is bliss. We didn't know we could do it, but we got it done in 10 years," he said, referring to the Apollo program. "Well, yeah, you were lucky," Mazzini said. "Now, you can't build a fighter jet and get an initial operating capability in 15 or even 20 years."

They then discussed the government's influence on defense and aviation. "There's so much revolving door between government and the industry that the defense companies are as inefficient as the government is," Mazzini said. Lash agreed. "I mean, if a defense contractor were to produce a cell phone, it'd be the size of this thing here," Mazzini said, pointing to a big rusted box. "Yeah. We used to have a joke about that," Lash said. "It went like the government wanting a wheelbarrow. By the time the functions were added, and all the organizations got involved, you got this huge device that isn't anything like a wheelbarrow."

Earlin Ward, a peppy engineer, joined the conversation. "We'd get so involved with everything that we'd lose the basic concept and concentration on the problem," Ward said. Even systems engineers sometimes end up sacrificing the big picture for details. A maze of procedures can overshadow the objective. "We got so much into writing the requirements, and adding more and more requirements, that we could never get to a fricking design. We couldn't do it," Mazzini said.

That's why the blue-box trainer was different. "Ed Link didn't need diagrams or systems manuals or anything. He understood his objective, and he understood his resources. And, he understood what capabilities he needed," Ward said. It was the *philosophy* of systems engineering Ed Link fully capitalized on. "This might all sound fairly simple, but go try to build that blue box. Some of that skills and substance are just not available anymore."

At the other end of the antiquary stood Gene Abbey, whose pioneering work with systems integration and testing on the Link simulators contributed to the success of the Apollo mission. With him was Bill Bennett, a retired software-systems engineer at Link, who worked on military flight simulators. They discussed a beam splitter that Abbey was fixing on the Apollo simulator on loan to Tech-Works! from the National Air and Space Museum.

Soon after delivering the Link Apollo simulator to NASA in the 1960s, Abbey built simulators for commercial nuclear power plants. Given the risks involved with their work, the nuclear power industry was quick to adopt simulators. After a pause in his discussion with Bennett, Abbey recalled the 1979 disaster in the township of Londonderry, Pennsylvania, at a nuclear plant three miles downriver from Susquehanna. The namesake, the Three Mile Island accident, was one of the worst in history. "The operators didn't recognize the problem till it got out of hand, and then it was too late," Abbey said. "If they had had adequate simulator training, they could have recognized and circumvented that catastrophe."

In Bennett's view, the Three Mile Island case was a scenario with no single output or outcome. "These are not simple calculations to do in your head," he said, remembering his programming for a portion of the Apollo simulator. "The only way you know is the whole thing comes apart." The software had to communicate with 1,500 circuits, each with switching and control functions. "The key was to embrace levels of abstractions and visualize these complex interactions in your mind's eye," Bennett added in his lulling voice. "That's precisely what Ed Link did."

When Bennett started working for Link in 1952, he had heard the company did systems engineering. "But what's that?" Bennett wondered. "It seemed like a word without a definition, you know?" He wasn't the only one who didn't initially understand systems engineering. Jim James, another retired engineer with a long, white biker mustache, echoed his sentiments. "The simulator could be the system with subsystems under that, or the simulator might be a subsystem under a complete training program," James said. "I mean, you know, it's all hierarchical."

In the early 1990s, Bennett wrote a monograph called *Visualizing Software*. He used the example of a simple road map presented in two parts. In the first part, only the dots representing towns are printed with their names, and in the second part, the lines represent the roads that go through the towns. This exercise requires a viewer to simultaneously look at two maps with different sets of information and synthesize them. The key was combining the maps into one mental image—with town names and road names—to see their interactions.

Humans navigate complex sets of information at many levels, prioritizing some details in one instance and another set in another.

Applying this idea to simulation software, Bennett argued for hierarchical design. "Most of the time, it's impractical to show a large piece of software as one single diagram," Bennett wrote. "Some means of breaking up a large software system into many diagrams, like the hierarchical division of the road maps, is essential." This basic multilevel approach underpinned the software development for the Apollo simulator.

MAINTENANCE IS A WAY of life for the seasoned engineers volunteering at TechWorks! This cohort of engineers views retirement as a time for restoration in addition to recreation. It's easy to forget the humdrum yet essential duties of repair, renewal, and reconstruction when we are constantly fed dazzling narratives about tech innovation. "Another flaw in the human character is that everybody wants to build and nobody wants to do maintenance," Vonnegut has written in *Hocus Pocus*. Scholar Debbie Chachra, in a nod to novelist Arthur C. Clarke, has said: "Any sufficiently advanced neglect is indistinguishable from malice." When we valorize moonshots, we might deemphasize maintenance. The former moonshot engineers at TechWorks! are examples of those doing unthanked, unglamorous, unrewarded, and unrecognized care work. Responsible engineers know that maintenance and modernity go hand in hand, not one at the expense of the other. Maintenance is the mother of modernity.

"To say we are underfunded is an understatement," Sherwood said, reflecting a common problem of nonprofits, particularly those committed to stewardship of older technologies. Bob Swarts is another volunteer, who revitalized a general aviation trainer with Carl Mazzini and others. The GAT-1 was Link's fully electronic, fiberglass system from the 1960s—a lightweight, fixed-base, plug-in machine resembling a single-engine aircraft, flying up to 10,000 feet. The restored trainer reproduced the cabin sounds accurately, and Swarts wanted me to take it for a ride.

I climbed, turned, glided, and cut wide arcs through a 360-degree turn as if grooving to Bob Marley reggae. Then, before I could think, I was diving and spinning. The trainer was shrieking. "Watch out. You are stalling," Swarts said. It was clear I had violated every known rule of flying. I stepped off the trainer with adrenaline and sweat as if I had landed on a tilled cornfield.

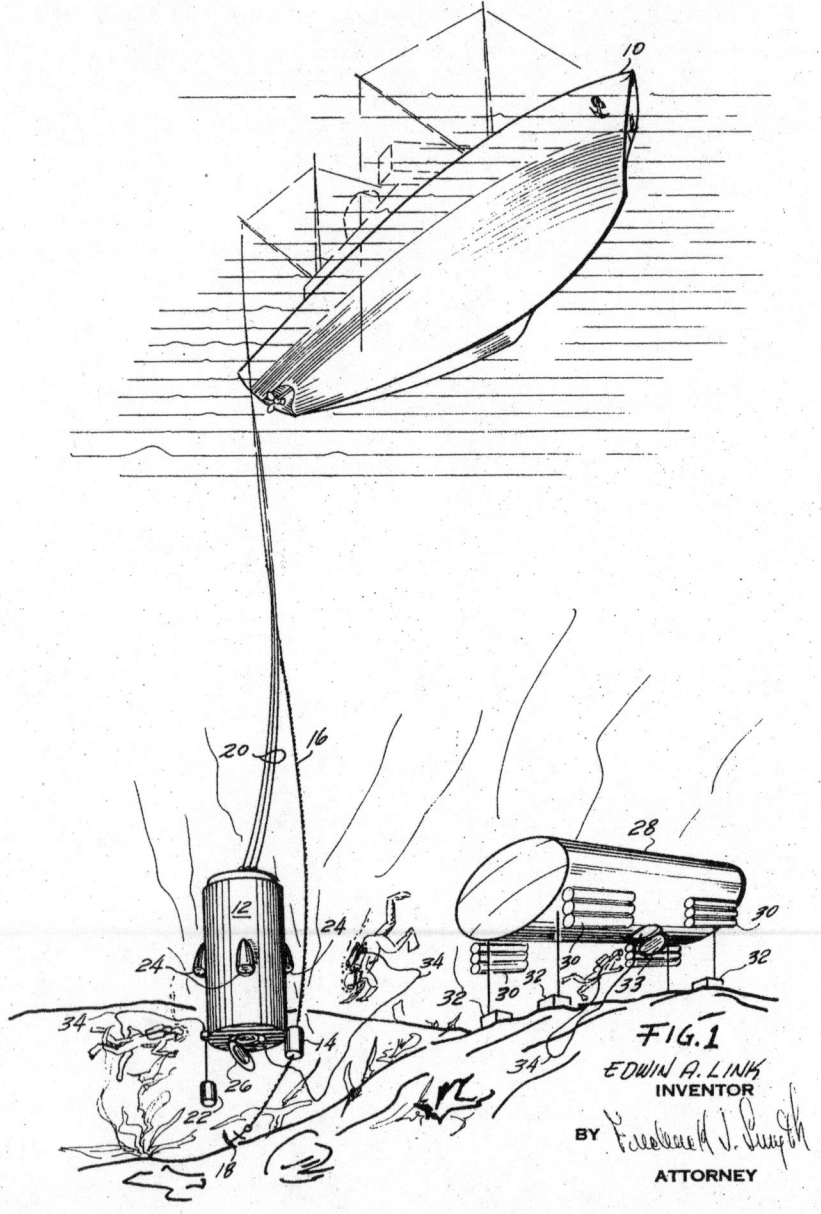

FIG.1

EDWIN A. LINK
INVENTOR

BY

ATTORNEY

Chapter 5

Sway

A 1957 PROFILE OF ED LINK IN *SPORTS ILLUSTRATED* described him as "part fish and part fowl . . . a restless sort of aquatic bird with a habit of disappearing frequently into the sky or sea." Ed Link had spent a career with his gaze fixed on the heavens, but his interest in the sea was relatively new.

To offset the stress of his wartime duties with the blue-box trainer, Ed Link took up fishing in the remote wilderness during the 1940s. He flew his Grumman Widgeon floatplane to the Great Lakes and the Canadian bush country during his off time. To solve the problem of transporting an unwieldy fishing boat on a small floatplane, he rigged up a collapsible canoe that could be disassembled and fit into two large suitcases. The sections could be reassembled into a watertight, functional canoe. Ed Link called the commercial version the Linkanoe, and it sold in the thousands. Evidently, he was ready for new engineering challenges.

Over the years, Ed Link's hobby evolved from weekend fishing in lakes and rivers to weeks-long excursions in the sapphire waters of Florida, the Bahamas, and Cuba. Shortly after World War II, he purchased the *Blue Heron*, a 43-foot yacht, to hone his sailing skills. Ed Link was able to transfer his navigational expertise from flying to sailing, melding information from navigation charts, radio guidance, and local weather forecasts. But as a newcomer, he was less bound to seafaring traditions. Ed Link shocked the judges in a St.

Petersburg–Havana race with an unconventional technique. When all the other competitors went south, he sailed west—and won.

Once Ed Link sold his aviation company, he had more time to focus on sailing. With his wife, Marion, and their sons, William and Clayton, as exploration partners, sailing became a beloved pursuit for the whole Link family. Although some could have seen this as a drastic change of focus, applying his expertise to a different context was a welcome challenge for Ed Link. He once said about his switch from the skies to the seas, "water and air are both fluids. One is just thicker than the other." He even designed a submarine to look like a helicopter. While rockets fired off from Cape Canaveral, just to the south near the Florida Straits, Ed Link mused about the underwater frontiers.

IN JUNE 1951, Ed and Marion Link accepted a friend's invitation to explore shipwrecks in the Looe Key of Florida, named after a small 18th-century British warship grounded in the War of Jenkins' Ear. Accompanied by professional divers and marine archaeologists, Ed Link plunged into the museum of the ocean floor to recover part of the wreckage, made of ivory. Hours later, he helped raise a heavy cannon from 1617. The experience solidified Ed's conviction: he would be a treasure hunter. "I took up golf once and put it back down very gently," Ed Link recalled later. "I don't like to get my hands dirty, but I like to get them greasy. Wreck hunting seems to have the right challenge."

Soon after his first shipwreck expedition, Ed Link convinced Marion to learn to dive. He wore an aqualung and fastened the mask over Marion. Then they slowly dipped into the crystalline world beneath the waves. "A beautiful sea garden stretched out on every side, a world of waving sea grasses, fantastic coral formations and lacy traceries of sea fans," Marion wrote. "The spiny creatures frightened me, and I backed away, lost my precarious balance and wavered helplessly toward another pit, which was liberally dotted with more sea urchins." Ed and Marion Link were instantly addicted to the world of sunken relics. "When we made those first exploratory dives, we were as green as the moray eels which curled malevolently within the jagged coral reefs," Marion wrote. She had imagined the wrecks as "hoary hulls still intact upon the bottom, and swarming with fish—dangerous and predatory fish, writhing octopi and huge rays with flapping wings."

For Ed Link, the engineer, the dives were a technical challenge. He needed to upgrade the *Blue Heron*. He purchased a 65-foot shrimp yawl, which he modified and christened as *Sea Diver*. The capacious workboat became their second home. Under the tepid blue waters of Italian islands, Ed and Marion Link found ancient Greek pottery, collected jewelry and coins, and retrieved antique kitchen utensils and cannonballs. "I was able to swim about this strange undersea ravine, floating up the steep sides of the coral rocks, peeping into eerie chasms and then gliding down to examine a nearly buried cannon, scarcely distinguishable under its disguise of coral," Marion wrote. She spotted the anchor of a sunken warship, coral-covered and almost hidden under a jutting rock shelf, and swam through its colossal ring. The shank, half buried in the coral, weighed over a ton. "I hovered over it enthralled," she wrote. "I was no longer fearful of this strange environment. I was at last home on the bottom."

In 1953, Ed and Marion Link found a smoothbore cannon likely used by Christopher Columbus. Enthralled by this find, they searched for the wreck of the *Santa Maria*, which led them to question where exactly Columbus had landed when he "discovered" the New World in 1492. The Links partnered with Captain Philip Weems, Ed's previous collaborator on the Celestial Navigation Trainer, to probe the dominant narrative surrounding Columbus's voyage.

Ed Link studied Columbus's journals, explored the north coast of Haiti, and aerially retraced Columbus's route to America. "I found out that no one even knew what a league was in Columbus's day—it could have been one mile or three. It seemed to us the only way to prove anything was to sail the islands at the same speed we calculated that Columbus did, checking the time elapsed, as Columbus recorded it, against our own, and against the landmarks Columbus saw." Ed Link challenged the accepted theory of Columbus's first landfall. Writing with Marion, he offered that the "Great Navigator" did not land on the swampy San Salvador but on the Caicos Islands, 200 miles to the southeast. When the Links detailed their findings in *A New Theory on Columbus's Voyage through the Bahamas*, published by the Smithsonian Institution, they instantly sparked controversy among historians. Some dismissed the Links as amateurs or told them to stick with collecting ivory tusks and picking over long-sunk pirate ships. In the *New York Times*, one Pulitzer Prize–winning Columbus expert at Harvard scoffed at the Links' findings, ascribing them to a "sailor's imagination." Nevertheless, Ed Link stuck with his conclusion.

In 1956, the Links also found their way to Port Royal, a Jamaican port city, home to the indigenous Taino population and once notorious for the sins brought by its settlers: piracy, prostitution, and the plantation economy. In what some saw as a divine decree, in 1692, a tsunami struck this "wickedest" land and killed thousands. The Links were astonished that despite the repeated disasters that the city endured—fires, earthquakes, and hurricanes—Port Royal was described as "a place of spectacular beauty rising in majesty from the sandy sea bed." They had to see it for themselves. Five years after Hurricane Charlie, the Links' crew anchored close to Port Royal's reefs. The water was murky, with barely any visibility for their dives. Silt from streams and muck from latrines lining the shore flowed into waters infested by barracudas and sharks. "It took guts and persistence to descend day after day into these unpleasant waters," Marion Link wrote. They recovered scores of rum and drug bottles; many were preserved flawlessly with coral coating.

Ed Link's crew airlifted portions of the brick walls presumed to be the ruins of Fort James, a fortress built by the British during the

18th century to guard the city against French invasion. The divers expanded their search, lured by the prospect of finding chests of gold cakes and coins. After a day's search, they returned with only a copper medal, clay pipe, broken porcelain, and beer bottles. When a storm suspended their search, the divers rigged up a wire cage suspended by an airplane to catch and sift their recovered debris. A small sparkling object from the airlift prompted Ed Link to dive deeper, but he was disappointed to find that it was only recently minted Jamaican pennies. After a week of fruitless exploration, Ed Link's crew was disheartened and ready to give up. Then, on the final dive on the final day, they uncovered an artifact long hidden from view: a British cannon weighing over 5,200 pounds. Ed Link's extraordinary finds off the coast of Port Royal were momentous for underwater archaeology. Over the following decades, extensive missions led by others with insights from Ed Link's early excavations helped make the "catastrophic site" into an incomparable UNESCO World Heritage Site with universal cultural value. The Port Royal experience only heightened Ed and Marion Link's appetite for further expeditions.

In early 1957, Ed Link retired *Sea Diver* after a stretch of wild weather and waves in the Florida Keys wrecked its platform and broke its anchor chains. His 1959 upgrade was a custom-built hundred-foot *Sea Diver II* with big booms and winches. The 168-ton, double-hull, and double-diesel steel cruiser was capable of 10 knots. It had two electric generators, two high-pressured compressors, two radars, six cabins, two large freezers, a fireplace, a four-burner stove, a dishwasher, a washer-dryer, a machine shop, and the most advanced technologies found in ocean liners: radios, radar, sonar, loran, echo sounders, air hoses, and airlifts. The space between the hull walls became tanks for oil and water. The vessel's underwater viewing chamber was a portal for the camera-ready, neon-colored vistas. It had an on-deck decompression chamber and two cruising boats—the 19-foot *Reef Diver* and the 15-foot *Wee Diver*. A smaller, battery-operated submarine, *Power Diver*, also supported more elaborate expeditions.

In the summer of 1960, Ed Link steered *Sea Diver II* into waters off the Mediterranean coast near Tel Aviv to explore Caesarea, an ancient seaport sunk by an earthquake in 130 CE. Herod the Great constructed Caesarea in Roman-ruled Judea—later Syria Palaestina, then Palaestina Prima, and now Israel—over 12 years in honor of "Octavian" Caesar Augustus. When the Greek philosopher Apollonius of Tyana visited the port city, he described Caesarea as "excelling all others there in size and in laws, and in institutions and in the warlike virtues of ancestors, and still more in the arts and manners of peace." The boat's underwater searching devices got to work: a magnetometer to detect metals, a fathometer to locate debris, a forceful jet hose to vacuum away the detritus, and an underwater television camera to record the dives. Caesarea was an engineering spectacle with colossal shrines, citadels, hippo-dromes, watercourses, and a columned promenade. The city was also an important port for ships ferrying between the trade centers of Egypt and Lebanon. As the Book of Jonah describes, when storms struck, "the mariners were afraid; and every man cried out to his god, and threw the cargo that was in the ship into the sea, to lighten the load." The twisty tempests, too often, took down the ships themselves.

Relying on the writings of the first-century historian Flavius Josephus, Ed Link's archaeological expedition charted the size and scope of the ancient port. Using a combination of aerial maps, water levels, and underwater measurements, Ed Link recalculated the size of specific structures in Caesarea. He wondered if Josephus exaggerated the size of the stones or if the book translation was incorrect. Aided by his calculations, the probing dives from the *Reef Diver* recovered structural elements, including fallen pillars, large wooden beams, and marble columns, as well as long-hidden artifacts, including amphoras, commemorative coins, pieces of sculpture, and Roman and Byzantine lamps brought on trade vessels over the Sea of Galilee, the crossway of civilizations. The crew of 11 also retrieved two exquisite Roman cooking pots from the first century CE and primitive stone anchors from even earlier. Ed Link entrusted these relics to the Israeli government, and they were widely exhibited at museums.

These diving expeditions gave Ed Link a fresh perspective. There were worlds underwater that were yet untapped by human civilization. Treasures like the Port Royal cannon and the Caesarea ruins were found over months of laborious back-and-forth between the land and sea. But Ed Link wondered, what other artifacts would have emerged if teams could stay underwater for the entire expedition? In his mind, living at the great depths of oceans for extended periods was a similar challenge to human flight in the air. As with the early aviators, Ed Link's ambition presented many unknowns, but these were known unknowns.

THE NOTION OF UNDERWATER living has appeared in many different forms throughout history: from ancient Indian scriptures, in which the Hindu god Varuna resides in the seas and ascends to land on a crocodile, to myths about Alexander the Great in which he spent time underwater in a glass chamber, looking at a colossal sea monster that took days to pass by. Engineers have long sought to create submergence vessels inspired by the possibilities and mysteries of underwater living. In 1620, Dutch engineer Cornelis Drebbel demonstrated a leather-wrapped wooden submarine in the Thames for the British Navy. It was never deployed for combat. Then, in 1776, David and Ezra Bushnell, brothers from Connecticut, worked with clockmakers to develop a wooden submarine called the *Turtle*. Though the vessel sank in the American Revolution against British warships, future president Thomas Jefferson brought publicity to the effort in 1798. Two years later, Robert Fulton prototyped the practical submarine *Nautilus* for the French Admiralty. Fast-forward to 1934, when the first deep submergence was demonstrated in a bathysphere, a spherical diving chamber lowered to 3,000 feet in the Caribbean. Twenty years later, inventor Auguste Antoine Piccard designed the first bathyscaphe—a self-propelled deep diver—that reached 13,300 feet in the waters off Dakar, Senegal. That record was broken in 1960 at the Mariana Trench by Piccard's last bathyscaphe, *Trieste*, which reached almost 36,000 feet, a depth exceeding the peak of Mount Everest by over a mile.

As Ed Link's interests evolved beyond underwater archaeology, he believed he could engineer the means for humans to stay on the bottom of the ocean for extended periods. He imagined that underwater living would unleash new opportunities from agriculture to national security. But the technical challenges that water posed were daunting. Water weighs much more than air and exerts greater pressure. Hence, staying on the ocean floor for extended periods means battling harmful ambient pressures. One atmosphere of pressure (ATM) equals 14.7 pounds per square inch. At 33 feet underwater, one experiences 29.4 pounds of pressure, or 2 ATMs. Every additional 33 feet underwater adds an ATM, or 14.7 pounds of air pressure. So, at 66 feet, 2 ATMs of water pressure and 1 ATM of air pressure combine to produce 3 ATMs of ambient pressure. Once humans descend deeper underwater, the blood circulation resembles a highly pressurized soda can. Crack the can open too quickly, and the contents burst out. For this reason, decompression, the careful process that returns humans to surface atmospheric pressure, can take several days.

Ed Link realized that deep submergence was better achieved with a submersible decompression chamber. The decompression chambers aimed to provide divers a safe place to return to ambient pressure while deep underwater. The chambers were to prevent two deadly effects of diving: "getting bent" and "getting narc'd." The first complication, the bends, causes sharp pains and vertigo because of bubbles in the bloodstream, which, without decompression, can lead to paralysis and even death. The second complication, nitrogen narcosis, is a form of intoxication from absorbing excess inert gases while deep underwater. Ed Link designed a one-person aluminum cylinder that could be lowered from the ship and hauled up like an elevator to afford more safety and comfort during the lengthy decompression ordeal. During his Port Royal expedition, he first tested the chamber—11 feet long and about 40 inches in diameter. In 1961, Ed Link met the famed Jacques Cousteau in Monaco. They agreed on a joint venture to use the decompression chambers to transport people to Cousteau's proposed seafloor houses.

In August 1962, the 58-year-old Ed Link, saturating himself with a helium and oxygen gas mixture (heliox), entered the diving cylinder. He descended 60 feet below his ship, anchored near the French Riviera, and remained there for eight hours. His 21-year-old son, Clayton, delivered macaroni and cheese to a relaxing Ed Link inside the cylinder underwater. A month later, under Ed Link's supervision, the Belgian underwater archaeologist Robert Sténuit became the world's first aquanaut. Sténuit spent 26 hours in the submersible decompression chamber 200 feet below the *Sea Diver* off Villefranche. These preliminary demonstrations led to successful tests in batches of mice and a seven-month-old goat named Caroline, who was treated to a "big meal of cabbage" after spending 16 hours at 400 feet deep. The team was awarded a grant from the National Geographic Society, officially kick-starting Ed Link's "Man-in-Sea" program.

To bolster support for his project, Ed Link wrote a prospectus to the US Navy describing how his program could simultaneously advance physiological studies, instrumentation development, and operational testing in deep-sea trials. The initial phase of the Man-in-Sea program focused on tests at 400 feet, a depth beyond the capabilities of standard scuba gear. The emphasis was on engineering inflatable underwater suites called Submersible Portable Inflatable Dwellings. The SPID was a bottom-moored setup with breathing-support systems. Another design was an igloo-shaped enclosure for workers to perform undersea pipeline-repair work. While testing these suites, the decompression chamber served two functions: a conduit to the SPID and the IGLOO and a conveyance back to the *Sea Diver* for deck decompression.

During this time, Ed Link also created Ocean Systems, which quickly attracted the US Navy's interest. The company's vision was to enable humans to live and work at depths of 1,500 feet, opening up an underwater region the size of Africa for research and exploration. "We cannot afford to neglect any longer the great resources of the oceans with man living and working directly in its environments," Ed Link wrote. "Wake up America, we are on the verge of a new era." He had a reason for optimism, given his tremendous success with his underwater developments up to this point. But Ed Link's dreams of deep submergence were disrupted by a dark disaster in 1963.

ON THE MORNING OF April 10, after a flurry of commands and checks, the USS *Thresher* began descent for a safety-check dive 220 miles east of Cape Cod. With 3,500 tons and 3,000 silver-brazed piping joints, the nuclear-powered hunter-killer was acclaimed for its stealth and silence in the great depths. The submarine and its crew of 129 people reached its assigned depth. Fifteen minutes later, the troubles began, but the sub's emergency call to the rescue ship *Skylark* was garbled. *Thresher* had attained its crush depth, at which a watercraft collapses under pressure. Death and destruction were instantaneous. The navy's closest rescue equipment worked only to 800-odd feet and couldn't go deeper. They brought in the bathyscaphe *Trieste* for the deep search.

Finding the wreckage in the saline shadows of the North Atlantic was agonizing. The failure was mysterious and the following congressional inquiries contentious. What about the ship's 20-ton nuclear reactor and its long-term impacts on the oceans? And what about the safety of other active submarines? Admiral Hyman Rickover testified that the *Thresher* accident "should not be viewed solely as the result of failure of a specific braze, weld, system or component, but rather should be considered a consequence of the philosophy of design, construction and inspection, that has been permitted in our naval shipbuilding programs." He argued that recent advancements "may have forsaken the fundamentals of good engineering." Eventually, investigations revealed that metal failures from corroded pipes contributed to the submarine's destruction.

The US Navy created the Deep Submergence Systems Review Group, tasked with advising the development of rescue vehicles for submarine crews and planning for future missions. Ed Link was perhaps the most practically accomplished engineer on this panel, and he proposed ideas for lighter-weight submarines. With the weight savings, he said, "it could be possible to double the striking force as well as even increasing the safety of the crew and possibilities of going to greater depths than our Polaris submarines can go today."

Ed and Marion Link docked their *Sea Diver* at the Washington Navy Yard and began their yearlong work on the design. But after President Kennedy's assassination rocked the nation, the Deep Submergence Systems Review Group received little publicity. Their final report sank without a ripple. This policy experience was a departure point for Ed Link. He had a fuller view of the problem than others, not that different from the flight-training mindset only two decades before. His mission went deeper: to strengthen the foundations of ocean engineering.

JOE MACINNIS, a 26-year-old Canadian doctor, read Ed Link's Man-in-Sea articles in *National Geographic* and was determined to work for him. "I went to his office and knocked him over with my enthusiasm," MacInnis said. Ed Link hired him that day and committed to a fellowship to support MacInnis's training in diving medicine from the Link Foundation, established a decade earlier to support work in aviation and ocean engineering.

MacInnis asked Ed Link about the *Thresher* tragedy. "We should not have lost all those men," Ed Link told him. "It confirms how little we know about the depths of the ocean." Then, Ed Link took

MacInnis around the sun-warmed *Sea Diver*. "For our next dive, we're going to place a small station on the seafloor at 400 feet," Ed Link told MacInnis. "This chamber will carry two divers down to the station and return them to the surface. It's going to take a lot of hard work to get it done." Ed Link appointed MacInnis as the medical director for that project.

On an ultrablue morning in June 1964, Ed Link and Joe Mac-Innis stood on the *Sea Diver*, ready for a test near the Berry Islands of the Bahamas. They watched Robert Sténuit and Jon Lindbergh, the cave diver and son of the legendary aviator, suit up. Sténuit and Lindbergh entered the submersible decompression chamber. They were winched 430 feet underwater, where they intended to spend a record-setting 49 hours in the SPID. A long hose fed heliox gas to the eight-feet-by-four-feet rubber station. This blend of breathable gases rendered the divers' voices indecipherable on the intercom, akin to the high-pitched squeaks of Donald Duck. As a solution, the divers used Morse code to communicate from the SPID to the *Sea Diver*—similar to how instructors once spoke with pilots training inside the blue box. Sitting just inches apart, the divers wrote messages to each other, which Ed Link could read as he supervised them on the coarse closed-circuit television.

Sténuit and Lindbergh went on outings, making friends with an extroverted 200-pound grouper that hovered near the exit hatch. The divers enjoyed good meals: canned hash, lunch meats, bread, fruits, and a rich lamb stew for dinner. While the divers could withstand the extreme pressures, their cans of food looked like a truck had run over them. "Living in the depths, I have become a creature of the depths, adapted to their pressures," Sténuit wrote about this experience in *The Deepest Days*. "I have always found joy in dangers lucidly accepted and prudently overcome." A triumphant Ed Link wrote in *National Geographic*: "Our mission was accomplished. It proved that humans can adapt to the inhuman conditions of the deep sea."

By 1965, Ed Link further improved the design of his submersible decompression chamber. He contacted John Perry, a high-energy Palm Beach entrepreneur, former army aviator, and submarine inventor. As Perry's business partner, Ed Link sketched the world's first lockout submersible. The *Deep Diver* had 21 acrylic viewports and could cruise at three knots on battery power. It carried a crew of four in two separate chambers, the forward and lockout chambers. The forward chamber remained at atmospheric pressure, providing the pilot and an observer with a "shirt sleeve" environment, requiring no special clothing or equipment. The lockout chamber could be pressurized to match the depth outside. A floor hatch allowed the two divers to enter and exit the sub once on the seafloor, tethered to the sub by long umbilical-like cords that provided breathing and communication support. The dimly lit eight-and-a-half-ton submersible could stay underwater for 24 hours. Ed Link designed the *Deep Diver* for maximum maneuverability, adapting the central feature of his flight trainer. Designed to operate at around 1,300 feet, it could maneuver up, down, and sideward and rotate on its axis with six degrees of freedom. "Living in the sea is good movie talk but a hard way to live and it isn't necessary," Ed Link said. But with *Deep Diver*, crews could "do their work and get back to their sub for dry clothes, a charcoal steak and some rest."

To Ed Link, the *Deep Diver* was just the beginning for submersibles. By the mid-1980s, he wrote, when "it will be commonplace for man to enter and work freely in the seas, we can foresee elaborate engineering developments scattered upon the ocean floor—underwater tunnels such as that contemplated beneath the English Channel, networks of telephone cables and forests of oil derricks with interconnecting pipelines all available for ready servicing." Humans could "live in the sea for a month, even up to six months if necessary," he added. He imagined underwater hostels for enthusiastic skin divers to vacation in. He dubbed the aquatic rides "the Volkswagens of the deep," noting the operators of such vehicles would require the equivalent of flight training.

Months before Apollo 11, while Neil Armstrong, Buzz Aldrin,

and Michael Collins were training in the Link simulator for their unprecedented journey, Ed Link was setting a record of his own. He reached down 700 feet in the Tongue of the Ocean, a region of incredible depth in the Bahamas. In an intense, six-week expedition, Ed Link's crew completed over 40 dives and spent hours on the ocean floor near the Great Stirrup Cay studying exotic marine life yet to be named. In the twilight glow of the ocean floor, Ed Link's small steps, in their way, were making a giant step for humankind. But despite Link's contributions to deep-sea engineering, the world of submarines remained extremely treacherous.

IN 1968, FOUR NUCLEAR submarines disappeared. The *Skipjack*-class USS *Scorpion*, Israel's *Dakar*, France's *Minerve*, and the Soviet Union's *K-129* lost 315 crew members in total. The circumstances of each disappearance were different, but the exact causes of each disaster remain mysterious. At the time, very little was known about the properties of materials used to construct these submarines and how these materials behaved at extreme depths, temperatures, and in contact with one another. Following the four submarine disasters over a year, new research characterized submarine alloys' durability and susceptibility to failure at great depths. Ed Link built the *Deep Diver* from "plow steel," or carbon steel, a special heat-treated alloy that he knew to be highly durable. But the new research led Ed Link to the costly conclusion that he must decommission his vessel. These disasters also bolstered his recommendation that the navy panel designate ocean engineering as a formal field of study.

Despite the costly loss of the *Deep Diver*, Ed Link forged ahead and fabricated a lighter and highly maneuverable submersible. The 23-foot-long, 10-ton submersible featured panoramic views and two hulls: the forward hull, with a four-inch-thick, transparent, acrylic sphere like a helicopter, and the rear hull, which was a cylindrical pressure vessel made of marine-grade aluminum alloy. Each hull could fit two people. The sturdy submersible contained sonar, telephone, intercom, echo sounders, magnetometers, and Doppler navigation. It also featured wide-angled, low-light cameras and other critical life-support systems. Ed Link credited the project's financial backer, Seward Johnson Sr.—son of Robert Wood Johnson, the cofounder of Johnson & Johnson—in the submarine's name: the *Johnson-Sea-Link*, many of whose mechanical and electrical parts were made of scavenged airplane material. The first model was commissioned in 1971, operating at a depth of 1,000 feet and later upgraded to triple that depth. The sub was primarily suited for search and recovery, archaeological applications, and delicate sampling of tiny creatures. Fifteen years after its debut, Disney World's *The Living Seas* exhibit deemed this design "futuristic."

But soon, Ed Link endured a calamity very close to home.

JUNE 18, 1973, WAS unremarkable at first. Four men were ready to make a routine dive off the Florida Keys to collect research specimens on *Johnson-Sea-Link I* under Ed Link and Joe MacInnis's supervision. But when they reached the seafloor, the submersible's cables became fatally entangled with a scuttled World War II destroyer. Because the dive was intended to be brief, the crew was ill-equipped for long-term survival. The navy dive teams were summoned but arrived late. They never reached the sub due to the drag forces from strong surface currents acting on their air-supply hoses. The outwardly calm Ed Link on the *Sea Diver* blinked tears, keeping the walkie-talkie pressed against his ear. Finally, a torpedo-recovery vessel managed to snag the sub and winch it to the surface. The lives of the pilot and observer had been spared due to the thermal insulation provided by their acrylic hull. But the two men in the rear chamber weren't as fortunate. They were in their swim trunks, expecting only a brief dive, and the aluminum hull afforded them no protection from the cold. Due to hypothermia and the carbon dioxide buildup in the chamber, they slowly lost consciousness and died. One of them was Clayton Link, Ed Link's 31-year-old son.

MacInnis couldn't sleep that night. He went to a pub, double rum straight up. "At this moment, I hate the goddamed ocean," he wrote. "I hate its size and its force. It is a malevolent son of a bitch whose currents and cold have just taken a good man's lifetime of effort and compressed it into tragedy." Even in grief, Ed Link concluded that the navy's rescue measures were inadequate. "This loss need never have occurred," Ed Link wrote to the Coast Guard. "The sole remedy—rapid, effective submarine rescue systems— must be developed and kept in strategic locations."

Ed Link's response was to launch a new breed of remotely oper- ated vehicles called CORD, the Cabled Observation and Rescue Device. The CORD overcame the problems of previous rescue systems. Unlike the unwieldy hoses and cables that prevented the navy rescue team from reaching the sub in distress, CORD's slen- der umbilical cord cut through the fiercest currents. It also had an airplane-type joystick that offered more precise control and advanced instruments that helped rescuers "work blind" in the ocean depths. Once again, Ed Link's familiar techniques from the early days of blind flight could be applied to negotiate a foreign envi- ronment. "It made us realize how helpless we really were," Ed Link said, reflecting on that dreadful day. "But now we have developed five different types of rescue systems, and it couldn't happen again."

ON A CITRIC-WARM FEBRUARY morning in Florida, Marilyn Link asked me to follow her car. The 93-year-old half sister of Ed Link steered with the alertness of a race driver. "I'm demanding; a stickler for details," she once told a writer. Marilyn Link received her pilot's license in 1946 and earned proficiency entirely on the Link Trainer. She couldn't become a full-time pilot, with flying jobs rarely available for women. She had a graduate degree and held some teaching roles, an administrative position at the Smithsonian, and an executive job at the Hughes Aircraft Company. Ed and Marion Link were her constant mentors. When Ed Link started a research institute with Seward Johnson on an abandoned sand mine on Florida's east coast, Marilyn Link became the organization's first managing director.

Marilyn Link and I pulled into the parking lot of the Link Port, a former swamp, now home to the Harbor Branch Oceanographic Institute, part of Florida Atlantic University. Gabby Barbarite, a marine microbiologist and science communicator at Harbor Branch, received us at the entrance. Harbor Branch has about 200 scholars and students focused on technical topics such as finfish aquaculture, cancer-cell biology, physical-biogeochemical ocean observation, and modeling. Among them are scholars in ichthyology, benthic ecology, phytoplanktonology, and zooplanktonology, who have devoted their careers to improving the health and preservation of aquatic life. "Our seas are doomed unless we act quickly," Ed Link said when serving as Harbor Branch's director of engineering in the 1970s. "Pollution and depletion of the rich resources of the ocean bottom are taking their toll. Even with a complete turnabout in man's attitude . . . it will take many years to reestablish the essential underwater environment that will preserve the oceans for use of future generations."

Ed Link condemned the oil industry for its egregious pollution risks. The companies "haven't taken severe enough steps to provide backup systems" for oil spills and other ecological disasters. Ed Link leveled some of his strongest criticism against Operation CHASE, a controversial toxic waste disposal program administered by the US government until early 1970. Operation CHASE, an acronym for "Cut Holes and Sink 'Em," intentionally sank ships in different locations to dispose of unwanted munitions. Some of these ships carried deadly chemical weapons or nerve gas. Ed Link thundered that the government was taking a "devil of a chance" with this program. "There is ignorance at all levels of government of the problems of pollution."

At Harbor Branch, Barbarite explained the workings of aquaculture and highlighted the latest research on harmful algal blooms.

Then she opened a hangar-like room that held the *Johnson-Sea-Links I* and *II*. In their 40-year underwater service, the two submarines completed nearly 9,000 dives retrieving vital artifacts from engineering history. In 1977, they helped recover the wreck of the Civil War ironclad USS *Monitor*, the long-lost armored ship. In 1986, the *Johnson-Sea-Link*s helped locate scattered parts of the space shuttle *Challenger* after it exploded in flight and took seven lives. For the *Challenger* alone, they completed a total of 109 dives. "Were it not for CORD, the solid rocket booster pieces couldn't have been found quickly, the *Johnson-Sea-Link*s wouldn't have recovered the failed O-ring seal, and the space shuttle program would most likely have been discontinued," ocean engineer Andrew Clark said. "Ed Link not only had the foresight to recognize the need for a new field of education but the fortitude to actually launch it himself." Even before he knew anything about Ed Link, Clark wanted to be underwater. As a teenager, he read issues of *National Geographic* wherever he spotted them to monitor developments in undersea technology.

Clark, now in his 60s, calls himself the black sheep of his family. His mother was a nuclear radiologist, his father a nuclear physicist, and his three older brothers were National Merit Scholars, but Clark dropped out of high school. At 16, he left home and hitchhiked to Louisiana, seeking a job as a diver in offshore oilfields. In the early 1970s, he worked for Sun Oil Company on the platform supporting the first crewed oil-well completion on the seafloor. "In the offshore business, we were beginning to apply in practice all those methods envisioned and proposed by Ed Link's Man-in-Sea program," Clark said.

But the school of hard knocks could only take Clark so far. "Everything I saw out there I wanted to do, the older fellas would tell me, 'Son, you need college for that.'" After being rebuffed several times, Clark relented and returned home. His father patiently taught him calculus and physics evenings and weekends. Clark went to the Florida Atlantic University for his bachelor's and master's in ocean engineering, a program Ed Link helped create, before completing a doctorate at the University of Hawaii. He joined Harbor Branch in 1979 for a Link Foundation internship. A decade later, he followed Ed Link's footsteps by serving as the institute's engineering director.

Clark's work has ranged from developing large crewed and uncrewed vehicles to testing small instruments for exploring the oceans in three dimensions. He has designed and deployed seafloor sensor networks for monitoring tsunamis and seismic activity, and broadband satellite communications to transmit data and images in real time from maritime vessels, buoys, and platforms back to shore. "Few major fields of study can be so clearly traced back to one defining event. Even fewer can be traced back to the imagination and initiative of one remarkable individual," Clark wrote. "It wasn't Ed's nature to simply make a recommendation and then leave it to others to implement."

Since working as the medical director on the SPID missions, MacInnis has studied leadership in extreme environments, with Ed Link among his prominent case studies. "Ed Link was the essence of an alpha curiosity, an action-driven curiosity, a solution-seeking curiosity," he said. In the 1980s, MacInnis was part of a French-led expedition to the *Titanic*, whose wreckage lay 12,000 feet deep in the waters south of Newfoundland. His subsequent expedition in the 1990s filmed the search and inspired James Cameron's 1997 epic. And in 2012, MacInnis advised the Hollywood director on his seven-mile-deep solo mission into the Mariana Trench in the *Deepsea Challenger*, a "vertical torpedo" sub. Ed Link's template of courage and commitment, MacInnis said, "has guided me and kept me safe."

April 23, 1968 E. A. LINK 3,379,021

INHABITABLE UNDERWATER STRUCTURE

Filed Sept. 28, 1964

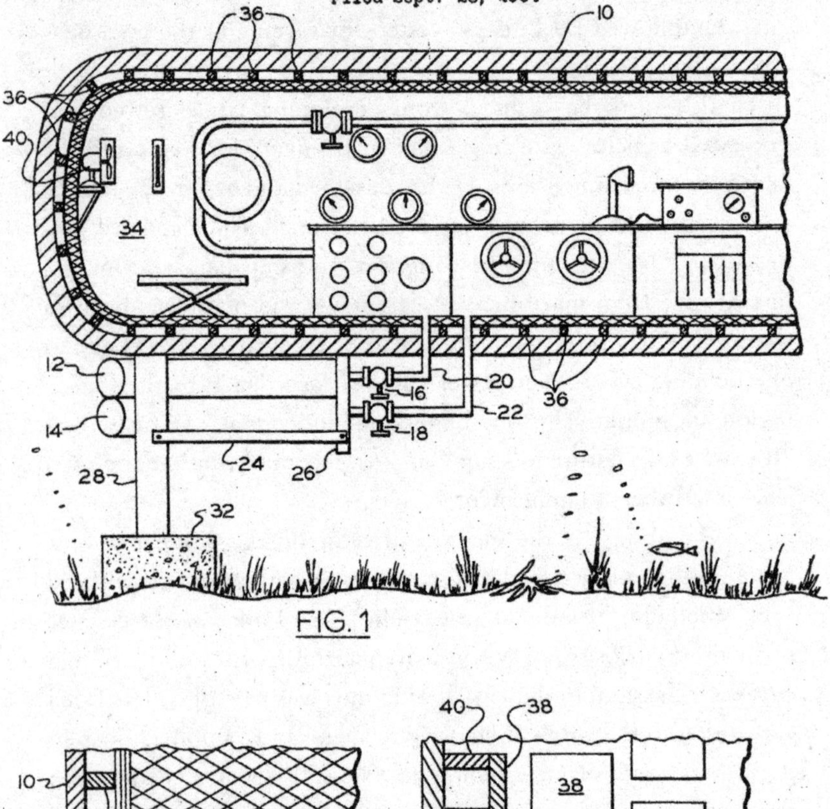

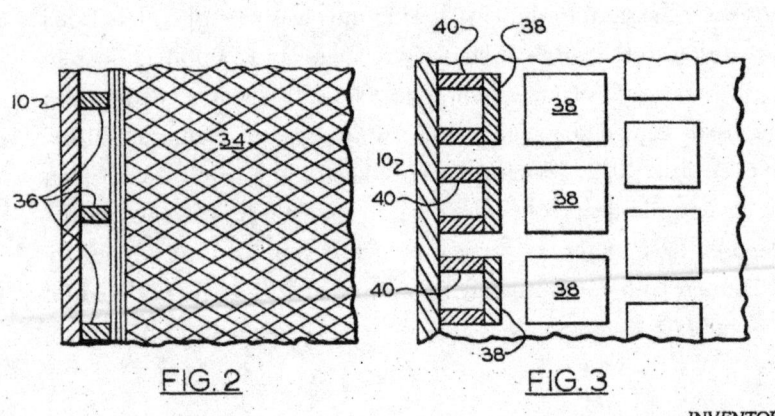

FIG. 1

FIG. 2

FIG. 3

INVENTOR.
EDWIN A. LINK

BY Frederick J. Burg, Jr.

ATTORNEY

Chapter 6

Heave

SIMULATED LIVING IS EVERYWHERE. WE "PROBABLY received simulators during the first days if not hours of our lives." Scholars William Moroney and Michael Lilienthal refer to pacifiers that serve as a *simulator* and a *stimulator*. "It simulates the nipple of a mother's breast or a feeding bottle and stimulates sucking and rooting reflexes, thus improving infant muscle tone." From pacifiers and model toys to imitation butter, simulated objects often script specific behaviors—nursing, running a toy train on tracks, or spreading butter on bread. Some early simulators were games that bonded entertainment with strategy. One example, *petteia*, a board game from fifth-century Greece, involved a strategic competition played with different-colored stones. In the sixth century in central India, the game *chaturanga* was developed. The name refers to the "four parts" of the Gupta dynasty—infantry, cavalry, elephantry, and chariotry. Future versions of this game, notably the *shatranj* of Iran, laid the foundations for modern chess, from which players developed tactical role-playing games that simulated potential conflicts. These *kriegsspiele*, or "wargames," would later ensure crucial victories on actual battlefields in the Franco-Prussian War. Wargames also had a long tradition in the United States, beginning in Naval War College under Theodore Roosevelt, then the assistant secretary of the navy.

In the early 20th century, wargames became routine, used to strategize the 1941 Pearl Harbor attack, the Nazi invasion of Belgium, and many campaigns during the Cold War decades later. Tabletop exercises called BOGSAT—an acronym for a "bunch of guys sat around the table"—have literally shaped human history. Wargames shrink an "approximation of war itself onto the surface of a table," notes scholar Jon Peterson. Because a wargame "was always just an approximation, it would never tame war, never control it—but it became an ideal vehicle for commanding, in the words of H.G. Wells, 'a game out of all proportion.'"

In 1974, Dungeons & Dragons was born, germinating a game genre. The description was simple: "Rules for Fantastic Medieval Wargames Campaigns, Playable with Paper and Pencil and Miniature Figures." The players could create their characters and act as their surrogates. The D&D quests entailed problem-solving, partnership-forming, treasure-seeking, and knowledge-creating. This form of role-playing "no longer simulated the experience of command—it simulated the experience of being a person who did many things other than commanding." The realistic but imaginative techniques of wargaming and role-playing that proved valuable in gaming the unthinkable would soon be applied to simulators.

ALTHOUGH QUITE UNLIKE LEWIS CARROLL, Jack Thorpe also believed in six impossible things before breakfast. Thorpe was a program manager for the Defense Advanced Research Projects Agency, or DARPA, whose singular mission is to imagine and invest in seemingly impossible technologies.

In the 1970s, Thorpe had his sights set on an area of national security needing transformation: military simulators. While simulators provided realistic practice scenarios, Thorpe contended that they did not recreate the intensity of close combat. As such, they were at risk of becoming rote. In 1978, Thorpe wrote a paper called "Future Views," presenting a 20-year vision for superior aircrew training. He proposed synching individual simulators that would plunge pilots into unscripted scenarios to rehearse for collaboration during combat. This approach was revolutionary: in Thorpe's vision, hundreds, perhaps thousands, of individual simulators would be "networked" to create a virtual world that participants entered and navigated together.

Thorpe was eager to transition from simulation devices to simulator networking. In SIMNET, trainees would no longer fight computers. Instead, teams of soldiers fought other teams of soldiers on the network. Each simulator within the network was stand-alone, with distributed controls and capabilities for real-time role-playing. For example, if one simulator was a tank, another served as a bomber, and a third was the command post. Although performing different functions within the simulation, each stand-alone simulator provided soldiers with the same highly realistic battlespaces to navigate. Many of the terrains for SIMNET were created from footage of real places—picture the opening combat on the beaches of Normandy in *Saving Private Ryan*. Each simulator came with its image generator and terrain data. They provided the operator with a unique line of sight, providing high contextual complexity and less display complexity.

Thorpe compared SIMNET to "Alice's looking glass" because it created a portal to the network that brought pilots, gunners, and drivers together as a team. Through this medium, individuals came together to hone their collective abilities and integrate their skills. "Instead of communicating with thick proposals or lengthy briefings," Thorpe argued, "government officials and legislators can live the weapon system in combat conditions." SIMNET's designers opted for selective fidelity rather than complete realism; the essential functions were included and the trimmings left out. The resulting prototypes of the simulated battlefields resembled early arcade games. Developers placed greater emphasis on the network's ability to train a team. Thorpe defined this focus on the network and deemphasis on realism as the "sixty-percent solution." It extracted maximum value from a 60 percent finished design, allowing for midcourse corrections rather than a completed design. Thorpe said, "Fast, approximate, and cheap was better than slow, deliberate, and expensive."

Eventually, SIMNET incorporated the kinds of panoramic graphics that we have come to expect from simulated environments. The system's graphics now recalled the visuals of action-adventure films, video games, and war itself. In a way, SIMNET's networked warfighting resembled a tryst between the Pentagon and Hollywood. Consider "flying carpet," a radical stealth visualization system on SIMNET. The technology swept its participants into the heat of combat. They could fly, zoom in and out on areas and persons of interest, and try god's view with the latest intelligence. The commander could customize and control the combat zone at will.

Modern warfare has integrated many aspects of simulation and become much more automatic. Advanced surveillance technologies the size of a housefly can provide an exacting eye-in-the-sky situational awareness to outwit adversaries. The Pentagon increasingly relies on autonomous weapons systems—devices like automated drones, robot sentries, and even "self-driving" submarines—to carry out hazardous missions. Soon, automation might infiltrate all combat elements, not just weaponry but decision-making, intelligence gathering, and commanding. But is this all "just war"? The feelings of moral abstraction or dissociation that these simulated guiltless environments might provoke in their "users" leave serious underlying questions unresolved. The ethics of remotely piloted drones and uncrewed aerial vehicles are not merely new concerns. These modes of war elicit unique feelings of inculpability, captured in the concept of "distant intimacy"—a hostility radically distanced yet close.

One SIMNET exercise involved reverse simulation—graphics reconstructing an actual event from accounts compiled after the fact. In early 1991, Thorpe's team recreated a panoramic battlefield of the Gulf War as though fought entirely in simulators. The "Battle of 73 Easting," named after the grid coordinates on the map, simulated combat in a blinding sandstorm. The programming synthesized information from actual battlefield tours, accounts of survivors, missile trails, satellite images, and army database reports about the explosions and casualties. This computer-generated exercise based on concrete reality trained soldiers' actions and emotions. Simulation now properly meant anticipation and, from it, an essential coherence. Scholar Patrick Crogan calls this "gameplay mode," in which the "reorientation is paired with a disorientation, just as reanimation is with a deanimation of what had come to life previously." The resulting rehearsal archetypes, such as common synthetic experiences or the joint simulation environment, enabled increasingly lifelike impressions for crew training and weapons testing.

"War is the province of uncertainty," General Carl von Clausewitz wrote. "Three-fourths of those things upon which action in war must be calculated are hidden more or less in the clouds of great uncertainty." The military philosopher understood that war was rife with psychological uncertainty and that no amount of technological exactitude—maps, graphs, or geometry—would simulate the true experience of war in a game. SIMNET vastly improved upon the early wargames and the stand-alone simulators of the 1970s. Yet, SIMNET posed a host of conundrums. In the words of one Hollywood producer, when simulators offer an environment that blends entertainment and education, "What's the difference between fighting Saddam Hussein or fighting Klingons?" Sometimes, even technologies with clocklike precision can come with cloudy moral challenges.

WHEN A 1973 OIL embargo increased fuel prices, the US military substantially reduced its open-air flight hours and substituted them with simulator hours. Flight simulators' sophistication and use grew throughout the decade, competing with the likes of aircraft, "light as a feather, though weighing billions of tons," in the words of Walt Whitman. Training on a simulator was not just safer, it was compellingly cost-effective. When the air force switched to a simulator, hourly training costs for one earlier bomber fell from $600 to $1. Consider the simulator for the B-52 Stratofortress bomber. It weighed 43 tons, with over 5,000 circuit boards and 13 computers. The 220 instruction manuals for the bomber alone weighed over 1,500 pounds. This simulator did everything a B-52 could: precision bomb sighting and following over 530,000 instructions. The only thing it didn't do was fly.

For the 50th anniversary of the invention of the Link Trainer, in 1979, Ed Link attended a commemorative conference at Piccadilly in London's West End. He and Marion Link visited all the major railway stations in the UK. They even ran a scaled-down steam-powered train through Harbor Branch Oceanographic Institute to show the potential of steam energy. Ed Link had been considering growing food for laboratory shrimps and oysters from sewage waste. He sketched a paddlewheel that used water to produce electricity.

One could imagine Ed Link retiring to his upriver Spanish-style stucco house with the operettas of Victor Herbert, regaling visitors with tales of barnstorming and underwater treasure hunting. But instead, after a beloved theater in Binghamton closed, Link purchased from the theater the pipe organ he had handmade and installed there in the 1920s. From then on, Link used his retirement to travel back to a largely forgotten period, filling his garage and small garden shack with a small organ factory. "I've had a fun career, I guess. First pipe organs, then aviation, then the bottom of the sea, and now I'm back to organs," he told a reporter.

Ed Link reignited his dormant love and resurrected the art of organ building that had given rise to the first Link Trainers. Professional organ builders were largely extinct, but Ed Link spent two years gathering the required parts. He sourced over 2,000 metal pipes from Germany, the Netherlands, and the US Northeast. Ed Link's financial adviser, Doug Johnson, remembered when the inventor turned up at the bank in his denim overalls with a suspicious request. Ed Link urgently wanted $10,000 in 20-dollar bills in a paper bag. He had located an original blue-box trainer in some barn and intended to repurchase it for parts to restore the organ. Ed Link assiduously revoiced the dusty, damaged pipes by adding new tubes for flexible tonal groupings from the Baroque and Romantic eras. The ensemble of precision controls, combination pistons, and wind-pressure adjustments produced a miniature symphony orchestra. The "string" instruments from the left side blended with the "wind" instruments of the right and unified with the center chimes and percussion.

True to his lifelong ethos of maintenance, when Ed Link complained that his wheelchair was poorly designed, he found the tools to rebuild it himself. His other passion, for his hometown of Binghamton, stayed with him to the very end. "Don't you ever believe any rumors that I'm moving anywhere. I love it here," he told a reporter. "I wouldn't live any other place."

Ed Link died in 1981, leaving behind an incomparable legacy of engineering.

THE SCOTTISH ENGINEER THOMAS TELFORD, nicknamed the Colossus of Roads, argued that engineering projects could have benefits beyond infrastructure, namely providing access to better education, jobs, agriculture, and health. He termed that process of building social capital through infrastructure a "working academy," through which "the moral habits of the great masses of working classes are changed."

In the 20th century, Binghamton had its own versions of the working academy. One such civic leader was George F. Johnson, a co-owner of Endicott-Johnson Shoes—or EJ. At its peak in the 1920s, the company employed over 20,000 workers and sold over 50 million pairs of shoes yearly. One 1934 advertisement bragged that these were "nifty shoes for the chap who is aiming to be somebody." The employees at EJ were predominantly European migrants who had landed on Ellis Island and gone straight to Binghamton. In return for their labor, EJ offered employees practically interest-free home loans, subsidized their food, and paid for their medical, dental, and childcare. Local philanthropy by the leaders at EJ soared, as did community pride. Wielding his progressive aims for the community, Johnson assailed the local chapter of the Ku Klux Klan active in 1920s Binghamton—a city once labeled the "Klan Capital of the North"—rooting the group out in a few years. The impetus was to improve the lives of his immigrant employees targeted by bigots and xenophobes.

Johnson liked to repeat that there were no rules in the company "except ordinary decency in human conduct." This ideal was part of EJ's Square Deal, which contended that company-wide profit sharing is as simple as "cutting melons or dividing plums." But this Square Deal was not purely altruistic. It was rooted in a "negotiated loyalty," whereby employers offered benefits to placate workers' demands. As historian Gerald Zahavi points out, EJ's welfare expenses were dubbed "efficiency expenses." In line with other welfare capitalists of his era, like George M. Pullman and Milton Hershey, Johnson cast himself as a "friend" of his workers. He sought to create orderly, wholesome communities to prevent any worker unrest. EJ Shoes' proactive paternalism fostered worker allegiance and stability, and profits. Indeed, this progressive industrialism made EJ and Binghamton a "veritable beehive of industry."

After learning about Ford Motor Company's successes and potential weaknesses, George Johnson organized EJ around his distinct community instinct. Ford's labor practices were "at a tremendous disadvantage," Johnson and his managers believed. "Their people do not live around the works. Mr. Ford does not live with the people—he goes into the works but seldom—they do not know him personally—it is all handed down to them through the medium of a lot of hired people—devoted people, good people, hard working people, but, still, hired people."

Between 1915 and 1950, EJ built thousands of houses for its workers. The "EJ Homes" designs followed a template, priced under $3,000. Johnson discouraged "EJ Families" from owning cars, persuading them to live within walking distance of the factories. He built parks, churches, and playgrounds in the triple cities of Binghamton, Johnson City, and Endicott. "There can be no security, there can be no guarantee of prosperity and industrial peace except through homes owned by the plain citizens," Johnson wrote in a letter urging a local banker to finance the homes. "I believe myself that the home is the answer to Bolshevism, Radicalism, Socialism, and all the other Isms," he added. "You will find that the home is the basis of all security."

In 1916, the company celebrated creating the 8-hour workday, the first major employer to reduce it from the standard 10 hours. During the Great Depression, EJ decreased his staff's working hours instead of furloughing them. Dubbed the "godfather of sports and recreation," Johnson made amusement and athletic facilities in the 1920s within walking distance of EJ homes. He gave pep talks to boost the energy and morale of his employees. And he shaped his identity as a "workingman's advocate" for relaxation and pleasure, encouraging weekend movies and horse racing. "The working people are no different than you and I and others," Johnson wrote to an associate. "We must be occupied, and we will be occupied." All this, of course, led to good business. In the summer of 1934, as a show of gratitude, over a thousand children, dressed in their best clothes and making festive music out of kitchen utensils, swarmed over Johnson's lawn to present him with a bouquet. Johnson had installed the sixth carousel for their enjoyment. The price of admission? One piece of litter.

Through the 1960s, EJ remained a dependable, respectable feature of the Binghamton community. "My father's family was an EJ family, and many of my aunts and uncles worked for the company," one resident recently said. "The really lucky ones got to work for IBM. That's how it was, you wanted to work for EJ's, but you wanted your kids to work for IBM." As this perspective suggests, Johnson's Square Deal would eventually lead to the Watson Way, and the focus of Binghamton's labor market would shift from the feet to the head.

THOMAS J. WATSON SR. didn't pay much heed to statistics. He preferred pithy, persuasive slogans. "Don't sell machines, sell results," he often reminded. "Emphasize applications, not hardware, the *why*, not the *how*."

Watson was born in 1874 in a small lumber-and-flour-mill town called Painted Post, an hour west of Binghamton. Keen in business, Watson sold sewing machines and pianos for a local family at country fairs. Then he joined National Cash Register, where, in 1911, he bellowed at a humdrum sales meeting: "The trouble with everyone of us is that we don't think enough." From that outburst came his soon-to-be-famous, one-word maxim: THINK. It defined not only Watson's identity but also the persona of his future megacorporation. This particular brand of simplicity and persuasiveness would one day make him America's Number One Salesman.

After he was fired from National Cash Register, Watson joined a holding company called the Computing-Tabulating-Recording Company in Endicott. A decade later, Watson renamed the company International Business Machines. IBM was first formed as an amalgamation of four firms, including the Bundy Manufacturing Company, renowned for its timekeeping devices. In 1892, one publication commended the Bundy time recorders: "No errors can be made in booking time, no disputes can occur as to accuracy, no jealousy is possible between time keepers and employees, neither can there be collusion between them." By the 1920s, time recorders became integral to the widespread efficiency movement of the era. Their underlying principles later informed IBM's "largest bookkeeping job" for the Social Security Administration. IBM's tabulating machines mutated into accounting machines.

IBM and its model of precision intelligence became a mindset. With the motto THINK, "I mean take everything into consideration. I refuse to make the sign more specific," Watson once explained. "If a man just sees THINK, he'll find out what I mean." The word was printed in giant block letters on the factory walls of IBM, emblazoned on coffee mugs, calendars, and concrete, and at one point, spelled out with red tulips on the company lawn. The steps leading to IBM Endicott welcomed visitors with the words "Think. Observe. Discuss. Listen. Read." IBM had its corporate songs and symphony orchestra. In one verse, choruses joined:

Our products now are known in every zone,
Our reputation sparkles like a gem!
We've fought our way through—and new
Fields we're sure to conquer too
For the EVER ONWARD IBM.

Watson's approach had undeniably religious tones. Company meetings in later years opened with a prayer delivered by an IBM pastor. People who didn't buy into Watson's spiritual enthusiasm for work were let go. For Watson, working at IBM meant a commitment to intense loyalty, as his biographers put it in *The Lengthening Shadow.* "Loyalty is the great lubricant of life. It saves the wear and tear of making daily decisions as to what is best to do," the IBM newspaper mused. "The man who is loyal to his work is not wrung nor perplexed by doubts, he sticks to the ship, and if the ship founders he goes down like a hero with colors flying at the masthead and the band playing." IBM emulated EJ's family spirit to boost employee loyalty and built a country club. In 1946, IBM helped found a college that became part of Binghamton University.

Like EJ Shoes, IBM believed itself to be the employee's friend, even if this friendship was one of unusual control and formality. Watson was always dressed in dark suits, starched white shirts with stiff collars, and striped ties, no matter the temperature. "He always shaved twice a day and bathed and changed his shirt at least as often," his biographers wrote. And like George Johnson, Watson loved the pomp of a good ceremony, chorus, and celebration. His staff, in return, displayed excessive esteem, once arranging an impromptu dinner for 2,000 employees. This "spontaneous trib- ute" included a colossal cake with 50 pounds of frosting. In the early 1950s, for Watson's 40th anniversary with IBM, the corpo- ration spent over a half-million dollars in pageantry, "giving three hundred and fifty tribute dinners in fifty-seven countries, with an attendance of well over fifty thousand." Such extravagances ended when IBM entered a new era under the leadership of Watson's son. With a more relaxed demeanor, Tom Watson Jr. was more of "a conservative rebel" who said IBM "needed wild ducks, not tame ones." And unlike his father, the younger Watson believed that loy- alty entailed "people caring about the company because it was their company, and the company caring about them."

BECAUSE OF HIS FAMILIARITY with EJ and IBM as a Bingham- ton resident, Ed Link also believed that loyalty to his company was "one of the finest experiences" of his life. "We don't have to sim- ulate cooperation," a 1960s Link company newspaper ad said. "In Broome County we have the real thing." In the late 1940s, a Link employee visiting England for business recounted his experience. "Many English workers have never seen their managing directors," he wrote. But this lineage of Binghamton leaders, including John- son, Watson, and Ed Link, made it a personal priority to be atten- tive to their employees and communities.

"Many employers like EJ, IBM, and Link enriched the commu- nity with their cradle-to-grave benevolence," said Tom Kelly, the retired business school dean and vice president of Binghamton Uni- versity. Kelly suggested that before corporate social responsibility became trendy in business schools, Johnson and Watson pioneered its earliest forms in Binghamton.

"When the nation got pneumonia in the 1930s depression, Binghamton got only a slight cold," Kelly said. By the end of the

1960s, the Binghamton region was home to some 200 firms, from the world's largest furniture producer to the oldest photographic supplier, as well as suppliers of clothing and cosmetics. Subsequent decades saw defense contractors moving into the area, and these strong trends continued into the 1980s.

On a crisp September afternoon in 1984, President Ronald Reagan visited Binghamton. He celebrated the "Valley of Opportunity," which represented, in many ways, America's story. "The computer revolution that so many of you helped to start promises to change life on Earth more profoundly than the Industrial Revolution of a century ago," Reagan said to a cheering crowd in an outdoor school stadium. His dream for America was "to see the kind of success stories in this valley multiply a million times over," he told the thousands gathered. "When those immigrants came to our shores and said, 'Which way EJ?' they were asking which way opportunity, which way peace, which way freedom."

In the decades that followed, Binghamton's experience belied Reagan's optimism. The region saw a tremendous population outflow and some of New York State's lowest economic growth rates. Jobs moved offshore, layoffs followed, factory buildings lay vacant, and property values plummeted. Once an industry capital and a role model for community building, Binghamton has faced stiff economic headwinds. In a 2012 well-being survey, it ranked second among the most overweight cities nationwide.

Baby boomers imagined working for one company their entire life with the expectation of a comfortable retirement. "They were confident that their children, my generation, could go even further," writes Kristina Wilcox in her study of the factory town. "Many saw these dreams, expectations—even 'givens' crumble before their very eyes."

IBM had 1,300 workers in 1914, then grew almost 10-fold by 1960. By 2008, the number of IBM employees grew globally by 30-fold. But all the while, the Endicott plant was being downsized. By 2014, the once-prominent buildings of the community were abandoned. The IBM Country Club was deserted; it became a victim of vandalism and was described as one of the area's "biggest eyesores." Universities and hospitals have replaced IBM as anchoring institutions.

Kelly suggests a cycle or pattern that has existed since Binghamton's river hub made colonial trading possible. "There was a kind of inspiration and inventiveness that enabled Binghamton's progress," Kelly said. "That led to new types of industries, new types of innovations, and new prosperities." As pressures of globalization adversely affect many American industry towns, Binghamton's opportunity, Kelly said, is to do what Ed Link did industriously well—routinely inventing and reinventing himself. One lesson from Binghamton's illustrious and ill-fated history is clear: prosperity is not permanent. "Remember: the past won't fit into memory without something left over," the poet Joseph Brodsky wrote; "it must have a future."

Not long ago, about 90 former employees of EJ got together. Some traveled long distances to Endicott and brought memorabilia from the company. "We kind of grew up together," one attendee, who worked for the company for 26 years, told the *Binghamton Press & Sun-Bulletin*. "It sounds trite, but it really was like family."

TOM KELLY AND I met during a busy lunch hour at the Lost Dog Café, a vibrant eatery in downtown Binghamton housed in a former cigar factory. Our view included the historic bluestone Victorian Gothic church, which rose into the bright June sky. Kelly recalled his own life experiences with simulation growing up in northeastern Pennsylvania. During his school years, students performed air-raid drills and practiced duck-and-cover in case of a foreign attack. The Binghamton region had been one of the prime targets in the United States during World War II and the Cold War, with all the espionage work underway, especially IBM's work for NASA.

In his recollections of Binghamton, Kelly lamented that companies don't make the kinds of long-term investments into the community that they once did. Even latter-day GE would be affectionately called "Generous Electric" for its investment in employee development in Binghamton. "Corporations are renting the talents of young people and putting them to work for short periods," he said. "I think it's a problem. We'll have to develop new kinds of lifetime learning and corporations enabling that." The industry loyalty culture that profited Binghamton now seems a vintage idea. Job-hopping and the gig economy have given rise to quicksilver career development, and community development has become less of a priority for employers and, perhaps, employees. Some may wish to invest themselves in the community but can't because of the existing job market. And some of these trends may point to a generational issue, or maybe not, but fickleness has displaced fidelity.

A block away from the Lost Dog Café is a one-time movie theater promoted as "absolutely fireproof," now converted to a brewery. Across from it is an old department store. Kelly and I walked into the adjacent dreary parking garage—concrete-gray and dusty, as it might have been at the Link Piano and Organ Company, which once stood on this location. Dulled wall art, a neon green background with fluorescent, *Matrix*-type font, read: "Welcome to the Birthplace of Virtual Reality." We walked down the ramp to the lower level and saw a fading mural on a corner wall behind oil-stained parking spots. This was where clocks became business machines, and the pianos reached for the clouds. Big block letters read: "On this site, Ed Link invented the flight simulator which transformed how pilots learn to fly 1929–1934."

We stood in silence. Between the piano factory and the parking lot, it felt like *The Twilight Zone*. Or, more precisely, "a dimension of sound, a dimension of sight, a dimension of mind," as the Binghamton boy Rod Serling usually opened his blockbuster series with the iconic four-note guitar riff. "You're moving into a land of both shadow and substance, of things and ideas." The garage ceiling was fractured, and the ribbed concrete bars were exposed and rusted. The floor was grimy.

I DROVE OFF THE parking ramp onto the asphalt quilt of Water Street. An Art Deco bridge relayed me across the Chenango River. "$Cash Paid$," a billboard read, soliciting Victorian furniture, old signs and toys, soda makers, pedal cars, pottery, and glassware. Then I drove down Clinton Street, the Antique Row. In the 1940s, this street buzzed with stores and bars, as writer Ronald Capalaces has recounted. Conversations in Slovak, Polish, Russian, Lithuanian, and Yiddish blending together with Italian filled one's ears. Restaurants enticed customers with hot pies, spiedies, and banana splits and shakes. While waiting for their prescriptions, ladies in the drugstore sprayed Evening in Paris perfume samples. The adjacent EJ shoe showroom had a peep-down fluoroscope to look at the foot—the sales clerk threw out the kids who were naughty with it. A nearby theater showed silent matinees accompanied by the music of a Link pipe organ. And during the city-ordered blackouts during World War II, people "walked up and down sidewalks in the pitch dark with cigarettes hanging from their lips," Capalaces has written. "It looked like fireflies glowing in the night and the cold of winter with snow covering the ground, the dancing dots of burning cigarettes made it seem like summer in winter. It was magical."

Today, the Antique Row has become antique. As if on an immersive simulator ride into the past, I took a deep breath of the time that no longer existed. The street is punctuated with onion-domed churches, funeral homes, coin-operated laundries, and a hip vegan joint. The old encountered the new, or as photographer Berenice Abbott said, "the past jostling the present." One can still spot the four square EJ company-built homes across the west side of Binghamton. Making my way through tree-lined streets, I reached Binghamton's Recreation Park, or simply the Rec Park, a few blocks from where Ed Link lived most of his years. A couple were walking their dog on the soft green grass with shade from the oaks and maples. A father and son played baseball in the adjacent field as a student idly strummed his guitar on the park bench. A teenager skateboarded in the parking lot.

The park's centerpiece is a 1925 Allan Herschell carousel in a large cupola. George F. Johnson donated the machine as free entertainment. The carousels were built by immigrants—ornate, naturalistic scenes and a "playful storybook" quality, as Herschell preferred. During its prime, the carousel operated 2,000 rides on any summer day. Although now painted with scenes from *The Twilight Zone*, this merry-go-round still maintains its 60 original hand-carved horses galloping to the original Wurlitzer military band organ.

I sat on a gallant horse with its head up and flowing mane. The operator rang the bell. And slowly, the carousel spun to a tune from 1926, "Take in the Sun, Hang out the Moon." This carousel was a system of gears, motors, and hand-carved dreams and delight. While technological revolutions of many kinds were happening in Binghamton, the carousel was also a revolution, both lilting and literal.

The operator rang the bell. One journey finished, and the next began.

The carousel whirled again for the Beatles' "Here Comes the Sun."

I was set free.

ABOUT FOUR YEARS LATER. Interstate 81 North, crossing into New York's southern tier from Pennsylvania, combines Endless Mountains with endless maintenance. The industrial park outside Binghamton once housed the wartime Link Aviation Devices and its future owners. The Court Street corridor to downtown is lined with architectural works of the master builder Isaac Gale Perry from the mid-to-late 19th century: the New York Inebriate Asylum or "Castle on the Hill," the Phelps Mansion, the Broome

County Court House, and the Perry Block, with a distinctive cast-iron structure.

Near the aerodynamic Art Deco bus station rises an imposing carved marble building. It's named after Sylvester Andral Kilmer, a bald, flamboyant, gunslinger-mustached physician who developed the famous Swamp Root formula, an alcohol-rich herbal concoction of the late 1800s. The brew claimed remarkable cures for human ailments until the 1906 Pure Food and Drug Act curtailed its sales. A comb manufacturer in the 1890s excelled with efficient automation a few blocks from the Kilmer building. In 1904, they sold a record 17 million combs made from elk horns and cow bones at 10 cents apiece before Bakelite and celluloid materials won the day. Now the former factory is an electrical substation bordered by empty buildings. The 1900-built Lackawanna Train Station, almost lost to neglect, has been revived and is home to small businesses and the last remaining of four radio towers. The Nobel-winning engineer Guglielmo Marconi erected this 97-foot steel structure for telegraphy in 1913. He successfully tested wireless communication to trains moving at 60 miles per hour between Binghamton and Scranton.

Tom Kelly and I met for lunch after the lean years of the Covid pandemic. It was a cold-swept January day with a wintry mix ready to rush in. "How quickly things change!" Kelly said at the Little Venice Italian restaurant, with Armando Dellasanta's impressionism and "Funiculì, Funiculà" in the background. "The Binghamton area has once again reinvented itself." Kelly was referring to the improving city economy and the rising cultural diversity of students and young professionals in the area. Binghamton University's downtown incubator has helped spur new technology businesses. The state and federal governments have invested significantly in a new, sizable lithium-ion battery manufacturing plant, where some of the original IBM buildings once stood, to create thousands of jobs. A new three-mile walking and biking trail stretched between downtown Binghamton and the university, and the local mall was turned into a "commons" with a sporting megastore and multi-purpose venue. And that dreary garage near the department store, once home to the Link Piano Company? It was razed, paving the way for a new parking garage.

I went to the Gorgeous Washington Street, where I lived as a graduate student in an apartment above a tattoo and piercing studio. The building next to it, once the historic headquarters of the Votes for Women Club, which led the local suffrage movement, is now a Chinese take-out joint. Past more restaurants, student housing, and the city's entertainment arena is the university's downtown center. The building's archaeological exhibit, *Our Invisible Past*, takes one to a different time. "Daily life is not usually the subject of written history, which tends to focus on significant events and prominent people," the description read. "However, everyone contributes in some way to a community's history." Binghamton is no single thing; it's simultaneously the bedrock, the people and the buildings that have come and gone, what it is now, and what it might become. The historic lenticular truss bridge a couple of blocks south, spanning the Susquehanna, was slick with frost. The South Mountain and Ingraham Hills appeared monochrome, starkly serene. I stood at Confluence Park, imagining the first communities that settled here. The Chenango splashed with the Susquehanna, quivering in the embrace.

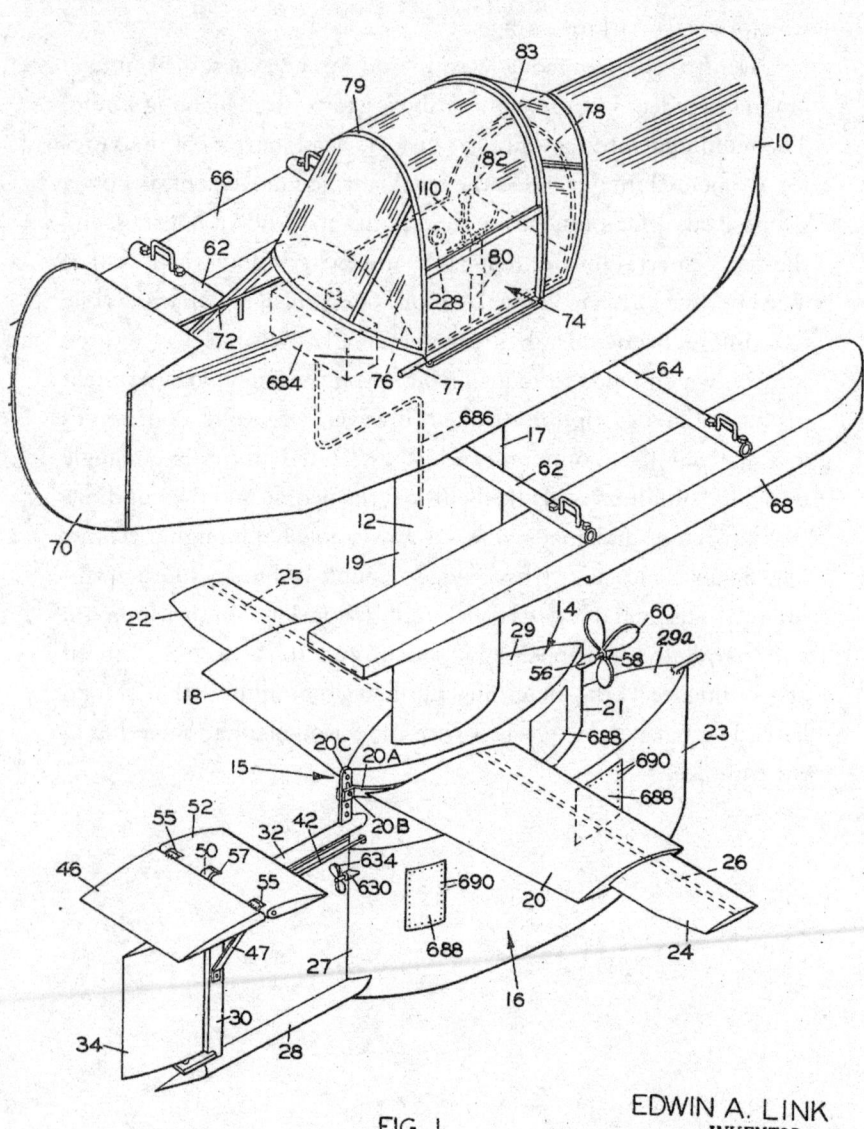

FIG. 1

EDWIN A. LINK
INVENTOR.

BY *Donald T. Hellier*
Philip L. Hopkins
ATTORNEYS

Epilogue

Thinking Inside the Box

> The many fictional inner journeys available to us—those that
> unfold in imaginary places—also come equipped with maps.
>
> —*Margaret Atwood*

THE EDWIN A. LINK FIELD, A FORMER COW PASTURE ON a Scottish settlement, is home to the Greater Binghamton Airport. With the heyday hurly-burly long gone, the airport is a small operation. The parking lot was vacant and grim in late autumn as distant lightning zigzagged. Muffled thunderclaps brought a brief rain shower. Airline counters were empty, with no humans in sight. The flight schedule on the electronic display showed two round-trip services a day. Automated security announcements overhead blended with an FM pop mix.

An easy-to-miss corner houses a display of Ed Link's Pilot Maker. After a lifetime of hectic hissing and huffing, it now stood with an air of reassurance, like a commanding bumblebee. The wingless wonder didn't describe knowledge; it deftly transferred it, committed to the idea that no single interpretation of the problem will suffice. And under the hood was an intimate grandeur of darkness that unified the real and the plausibly real.

If Karl Popper formalized a distinction between clocks and clouds, Ed Link nimbly coupled them without leaving the ground, one forming and informing the other. The Link Trainer work was far ahead of today's deep-learning devices. Yet, as a "depth technology," Ed Link's blue box recognized multiple dimensions related to one another. Perceptions become renditions only through relationships. As scholar Judith Roof has noted, "It always takes at least two—two eyes, two objects, two fields, two images."

Modern headsets and goggles now connect us with immersive worlds that span everything from concerts to competitive gaming. Even our politics operates in virtual worlds, concerned more with perceptions than facts. "It exists for only the fleeting historical moment, in a magical movie of sorts, a never-ending and infinitely revisable docudrama," one observer put it. "Strangely, the faithful understand that the movie is not true—yet also maintain that it is the only truth that really matters." In Ed Link's engineering, though, virtual reality conveyed an operational mentality, a code of conduct. Nothing looked more commonplace than this minimalist masterwork. It's a cage to liberate us from our cognitive cages.

"GREAT EVENTS ARE PERHAPS so only for small minds," the French poet and critic Paul Valéry wrote. "For more attentive minds, it is the unnoticed, continual events that count." Special events like the Wright brothers' first flight or the Apollo 11 landing are celebrated in the history of flying machines. They are headline episodes in the annals of aviation. However, the engineering revolution that Ed Link fomented in relative obscurity deserves greater appreciation. The Link Trainer democratized an activity previously available only to an adventurous few. It enabled broader participation for women in an arena defined by men's interests, giving those women vital roles of instruction, obscured yet integral.

After Kitty Hawk, the Wright brothers rose to an exalted status. Some characterized them as "an instrument in the holy hands of God." They transformed the world by wresting monopoly over the skies from birds. But without leaving the ground, Ed Link was equally transformative in imitating that reality. His work remains a quiet triumph in a world of thunderous takeoffs and show-offs. The point is not to create arbitrary fame contests between the first practical flight and the first practical flight trainer. There's hardly any controversy over who engineered what, when, where, and how. But a more profound issue lurks: how society assigns fame, manufactures eminence, propagates popularity, and ultimately judges one individual over another. "Due recognition is not just a courtesy we owe people. It is a vital human need," writes scholar Charles Taylor, describing how our identity is partly shaped or mis-shaped by recognition, with its presence and absence. In a relent-less ratings-and-rankings culture, someone will always get more publicity than someone else, whether worthy or not. In competitive fame-and-fortune seeking, we discount the dull, diligent, dutiful, and deserving.

The synthetic capabilities enabled by the Link Trainer now underpin the entire aviation enterprise. Today, they have democ-ratized access to education, combining psychology, sociology, and technology to produce knowledgeable and skillful crews in many areas. Just consider the collective cost of World War II without flight trainers, the sheer feasibility of the original moonshot, the viabil-ity of the modern airline business, and even the global commercial and cultural connectivity. If the Wright Flyer was an extraordinary first, the Link Trainer was an even better first. Flying a prototype a few times to succeed in a 12-second demo project and making it last 59 seconds doesn't need systems engineering. Still, at scale, systems engineering can prevent costly consequences when many lives are at stake. The first flights were an experiment, but the Link Trainer cultivated responsibility. As Wilbur Wright himself put it, one can fly without motors but not without knowledge and skill. The train-ers afforded far more nuanced instructions on performing in the total environment than simply flying by the seat of your pants. The blue boxes and the instructors who wielded them provided critical systems engineering support for what had been missing in aviation practice. Like the Wright brothers, Ed Link had no academic bona fides. He was drawn to challenges throughout his lifetime, observed Ralph Flexman, an aviation engineer who knew Ed Link for three decades. "While he did not go about looking for problems, when he came upon a situation or condition that produced unacceptable consequences, it always attracted his attention," Flexman noted. "Ed never worried about the solution to a problem until he thoroughly understood the problem in all its dimensions."

Ed Link engineered unusual and initially unintuitive devices for unique environments without any formal education in those areas. "By missing a university education, I failed to learn my own limitations," he once said. "It made it possible for me to do things that had never been done before." His brilliance was in liberating himself from specialization that allowed him to transfer concepts from musical space to aerospace and then to hydrospace, and finally return to where he started.

"Ed Link is the most persistent son of a gun I ever met," Art "Silver Bar" McKee, a noted treasure diver, put it in 1957. "Look at this boat," he said. "On the way here the radar breaks. Ed flies a man in from Miami and then does 75% of the work himself. Out of Nassau an icebox line broke, so Ed repairs the whole icebox. At Great Inagua a bearing or something in the automatic pilot froze. Ed goes below and turns it down on the lathe. He made a new belt for the air conditioning. He made parts for the water pump and the main generator." Ed Link's direct experiences in hostile environments underpinned his engineering accomplishments. The resulting designs defied distances and depths and recognized the overlooked relationships among technology, behavior, and policy early on.

In Binghamton, Ed Link often blushed when people called him a genius. "No-o-o, no, I'm not," he told a reporter. "First, know what the problem is," Ed Link offered. "Anybody can do the same thing I do if they're willing to work at it and study it. . . . Most people don't want to spend that much time." On a different occasion, when asked how an invention works, Ed Link replied, "Define the problem." When the same journalist asked Ed Link about the age of specialization, he said he didn't regret it. Still, he warned against overspecialization that "can result in forgetting other people's problems." Ed Link didn't have a command of language—he spoke less in words and more with his eloquence in engineering. There was the testing of self in his approach, even stretching personal limits from one domain to another. Ed Link's systems design undeniably had a placid sense of simplicity, veracity, and gravity, applicable to solving, resolving, and dissolving hard, soft, and messy problems. "If you are going to build these things," he said, "you have to have the integrity to test them yourself."

DOWN THE HALL FROM the Link Trainer in the Binghamton airport was a faint wall-mounted exhibit on the city's evolution "from the smokestacks to high-tech." It contained a copy of *E-J Worker's Review* from 1920, Thomas J. Watson Sr.'s THINK desk plaque, and three generations of IBM technologies from Endicott—the vacuum-tube, solid-state, and solid-logic circuits. A nearby display promoted spices and marinades, featuring spiedie sauce, a mainstay of the summer state fair and the city's hot-air balloon rally.

You'll find some offices and an observation room on the airport's second floor. The long linear concourse to the north end of the building went past the desolate departure lounge, a barren baggage claim, and an old marketing phone board for hotels and taxis. Pillar advertisements promoted the university, philharmonic, hospitals, casino resorts, and local restaurants. A stairwell led to a well-lit panoramic viewing deck studded with potted plants.

The dreamlike stillness in the room felt like the hermitage of a quiet genius. Ed Link's mural, 8 feet by 20, was the centerpiece. Across from it, a small memory collection featured fading pictures of player pianos, blue-box production, the Apollo program, the Perry-Link *Deep Diver*, the *Johnson-Sea-Link* submersible, and modern simulators. A long airport bench looked out the glass wall. The jet bridges were retracted; and the asphalt runways and taxiways, wet and illuminated. The weather vanes responded to the thick breeze as cumulus, nimbus, and stratus rapidly remixed in silent commotion.

For the Sanskrit poet Kālidāsa, the original Romantic centuries before Shelley, Coleridge, and Wordsworth, cloudy inexactness hinted at more than their expressions, perhaps a lover's longing. The departing weather front thinned out glittery gray cauliflowers; aimless cotton puffs were airborne against a salmon-hue backdrop. The twilight shades were sharp as the sun sank behind the hills. A clock ticked in the background as light retired to darkness. Everything seemed calm; nothing, constant.

Acknowledgments

Tango hotel alpha november kilo sierra to editor Brendan Curry, the reliable copilot in my flights of fancy. Assistant editor Caroline Adams brought her keen and caring sense to my high-g manuscript maneuvers. She fastened the seat belt on bumpy sentences and showed the exit sign for many more. Copyeditor Sarah Johnson completed critical cross-checks, as Robert Byrne, Rebecca Homiski, Will Scarlett, Steve Colca, Gabrielle Nugent, Lauren Abbate, Rebecca Springer, Victoria Keown-Boyd, and others at W. W. Norton & Company created the ideal cabin conditions for a straight-and-level journey. My agent Michelle Tessler, ever diligent and reassuring at air-traffic control, ensured the project was safe and orderly from before liftoff to after landing.

It's my privilege and pleasure to work with gifted colleagues at the National Academy of Engineering and collaborators across the National Academies of Sciences, Engineering, and Medicine—an incomparable free-trade zone of intellect. My mentor and friend Norm Augustine is a fount of wit, wisdom, and worldly guidance: "To think outside the box, go ask someone who isn't in the box." Roger McCarthy smoke-tested, Zach Pirtle sanity-tested, and Jag Bhalla wind-tunnel-tested the drafts. Geeta Bhatt, Lindsay Ruel, Stephen Nichols, Tom Snitch, and Ryan Alimento bravely beta-tested the book.

Binghamton—the community and the university—served as a knowledge base camp to unify my technical, business, and civic leanings. David Sloan Wilson and Tom Kelly have been inspiring "smart grids" of acumen and altruism since I was a graduate student. I thank Shelley Dionne, Doug Johnson, Andy Clark, trustees, and staff of the Link Foundation, and importantly the late Marilyn Link, for their encouragement. I applaud the vital service of Gerald Smith, Brian Frey, Broome County Historical Society, Roberson Museum and Science Center, and TechWorks! in communicating Binghamton's illustrious history. My thanks to Frank Cardullo and other instructors of the flight-simulation course at Binghamton University organized in cooperation with the American Institute of Aeronautics and Astronautics: Olen Atkins, James Davis, Thomas Galloway, Valerie Gawron, David Gingras, Jeffrey Schroeder, James Takats, and Andreas Tolk. Beth Kilmarx provided significant guidance during my research in the Bartle Library archives, as did the special collections curator Diane Newman and the late oceanographer George Maul at the Florida Institute of Technology. I am grateful to Wanda Brogdon and Julianna Andrews for their extensive assistance at the Library of Congress.

Longtime friend Barb Oakley has been a considerate "first responder" to my ideas. And finally, my boundless gratitude is for my parents and family. I present this book in memory of my father-in-law Sreekrishna Ramaswami. A steady and sagacious guide, he encouraged and engaged with this book before and unlike anyone else did. My wife Ramya's trusted, talismanic presence through the world's treats and turbulences exemplifies philosopher Simone Weil's prudence: "Attention is the rarest and purest form of generosity." She is my life's Link Trainer.

And DJH, you remain my beacon.

Sources and Resources

Collections

The British Library, London, United Kingdom
Broome County Public Library, Binghamton, New York
The Center for Technology & Innovation, Binghamton, New York
Edwin A. and Marion Clayton Link Collections, Glenn G. Bartle
    Library, Binghamton University, State University of New York,
    Binghamton, New York
Edwin A. and Marion C. Link Special Collection, John H. Evans Library,
    Florida Institute of Technology, Melbourne, Florida
Library of Congress, Washington, DC
Smithsonian National Air and Space Museum Library, Washington, DC

Notes

Prologue: Airplane Mode

1   "Say, it's only a paper moon": Music by Harold Arlen, lyrics by Edgar Yipsel Har-
    burg and Billy Rose (1932), popularized by Nat King Cole, Ella Fitzgerald, and
    Benny Goodman.

1   50 British cadets: Tom Killebrew, *The Royal Air Force in Texas: Training British
    Pilots in Terrell during World War II* (University of North Texas Press, 2003);
    Tom Killebrew, *The Royal Air Force in American Skies: The Seven British Flight
    Schools in the United States during World War II* (University of North Texas
    Press, 2015).

1   blue book: From *Notes for Your Guidance*, issued by the Royal Air Force, taken
    from Killebrew, *Training British Pilots*, 67; Davenport Steward, "As the English
    See Us," *Saturday Evening Post*, October 11, 1941.

2   500 novice students: Gilbert S. Guinn, "British Aircrew Training in the United
    States 1941–1945," *Air Power History* 42, no. 2 (1995): 11.

2   President Franklin Roosevelt: Mary Stuckey, *The Good Neighbor: Franklin D.
    Roosevelt and the Rhetoric of American Power* (Michigan State University Press,
    2013).

2   "arsenal of democracy": From Roosevelt's Fireside Chat, December 29, 1940, The
    American Presidency Project, University of California–Santa Barbara.

2   the lend-lease project: See R. G. D. Allen's "Mutual Aid between the U.S. and

the British Empire, 1941–45," *Journal of the Royal Statistical Society* 109, no. 3 (1946): 243–77; Warren F. Kimball, *The Most Unsordid Act: Lend-Lease, 1939–1941* (Johns Hopkins University Press, 1969).

2   **coat of arms:** Guinn, "British Aircrew Training," 6.

2   **corn on the cob and grits:** Killebrew, *Training British Pilots*, 83.

3   **segregation:** Killebrew, 68–69; Guinn, "British Aircrew Training," 16.

3   **Supermarine Spitfire . . . and Hawker Typhoon:** Killebrew, 84.

3   **"dust bowl":** Killebrew, 65.

3   **Stearman . . . "ground loop":** Killebrew, 30–31.

4   **landed his Stearman:** Killebrew, 76.

4   **Calvin Coolidge:** Ellen Pawlikowski, "Surviving the Peace: Lessons Learned from the Aircraft Industry in the 1920s and 1930s" (master's thesis, National Defense University, 1994), 2.

4   **Ed Link's pilot trainer:** Edward Molloy and Ernest Walter Knott, *The Link Trainer* (Chemical Publishing Company/Doray Press, 1941).

5   **"In a dark time":** Theodore Roethke, *Collected Poems of Theodore Roethke* (Doubleday, 1966), 239.

6   **"perfect all-weather operations":** Erik Conway, *Blind Landings: Low-Visibility Operations in American Aviation, 1918–1958* (Johns Hopkins University Press, 2006), 4.

7   **"were dismissed simply as":** James H. "Jimmy" Doolittle and Carroll V. Glines, *I Could Never Be So Lucky Again: An Autobiography* (Schiffer Publishing, 1995), 129.

7   **preparing a meal:** Stephen Darcy Chiabotti, "The Glorified Link: Flight Simulation and Reform in Air Force Undergraduate Pilot Training, 1967–1980" (PhD diss., Duke University, 1986), 3.

7   **half or more:** Reported in *New York Daily News*; Susan van Hoek and Marion Clayton Link, *From Sky to Sea: A Story of Edwin A. Link* (Best Publishing Company, 1993), 47.

8   **"An airway exists on the ground":** Quote by airmail pioneer Paul Henderson from Nick Komons, *Bonfires to Beacons: Federal Civil Aviation Policy under the Air Commerce Act, 1926–1938* (Federal Aviation Administration, 1978), 125.

8   **"doubtful that any piece":** Keith Matzinger, "That X*/#&** Link Trainer," *Aerospace Historian* 31, no. 2 (1984): 125.

8   **"passion for wings":** Robert Wohl, *A Passion for Wings: Aviation and the Western Imagination, 1908–1918* (Yale University Press, 1994).

Chapter 1: Pitch

11   **Dole wanted to produce:** Details of the Dole Derby are based on Lesley Forden, *Glory Gamblers: The Story of the Dole Race* (Nottingham Press, 1986); Jason Ryan, *Race to Hawaii: The 1927 Dole Air Derby and the Thrilling First Flights That Opened the Pacific* (Chicago Review Press, 2018). See also Jane Eshleman Conant, "Death Dogged the Dolebirds: Pioneer Pacific Fliers Wrote Tragic Chapter in Air History," *San Francisco Call-Bulletin*, October 10, 1955; Robert Hegenberger, " 'The Bird of Paradise': The Significance of the Hawaiian Flight of 1927," *Air Power History* 38, no. 2 (1991): 6–18; William G. "Burl" Burlingame, "The Dole Derby," *Honolulu Star-Bulletin*, December 29, 2003.

11   **"greatest air race" . . . Roman gladiators:** Ryan, *Race to Hawaii*, 217.

12   eight contenders: The ninth plane, *Spirit of Peoria*, was disqualified for insufficient fuel capacity.

12   Colonel Pineapple: Ryan, *Race to Hawaii*, 187.

12   "When we get over to Honolulu": Ryan, 216.

12   *El Encanto* was ready: "Descriptions of the Dole Derby Planes," *Aviation*, August 22, 1927, 414–15.

12   "I would rather have": Ryan, *Race to Hawaii*, 228.

13   "There it is": Forden, *Glory Gamblers*, 118.

13   "Let's go home": Forden, 104.

14   "For God's sake": Ryan, *Race to Hawaii*, 253.

14   One poet wrote: Addison N. Clark, "Into the West," *Western Flying Magazine*, 1927; Forden, *Glory Gamblers*, 137.

14   Newspapers declared: Quoted in Ryan, *Race to Hawaii*, 263.

15   The winning Art Goebel: Ryan, 170, 267.

15   Jensen was on a job: Ryan, 269.

15   "America has found": Scott A. Berg, *Lindbergh* (G. P. Putnam's Sons, 1998), 188.

15   The Binghamton region: William Lawyer, *Binghamton: Its Settlement, Growth and Development, and the Factors in Its History, 1800–1900* (Century Memorial Publishing Co., 1900); John H. VanGorden, *The Susquehanna Flows On* (Wilcox Press, 1966); J. B. Wilkinson, Tom Cawley, and John Hart, *The Annals of Binghamton of 1840: With an Appraisal, 1840–1967* (Broome County Historical Society and Old Onaquaga Historical Society, 1967); Ross McGuire and Nancy Grey Osterud, *Working Lives: Broome County, New York, 1800–1930* (Roberson Center for the Arts and Sciences, 1980); Gerald R. Smith, *Sweeping across America: Stories of Broome County Citizens in American History* (Keystone Digital Press, 2016).

15   "form a cabin": Lawyer, *Binghamton*, 5.

15   Bingham: See Margaret Brown, "William Bingham, Eighteenth Century Magnate," *Pennsylvania Magazine of History and Biography* 61, no. 4 (1937): 432–34.

15   Erie Canal: See Peter Bernstein, *Wedding of the Waters: The Erie Canal and the Making of a Great Nation* (W. W. Norton, 2005).

16   Factory furnaces: Mass production of cigars created an early wholesale factory model in Binghamton. Over 6,000 workers in 70 firms rolled over 150 million cigars.

16   Parlor City: Tom Cawley, "Binghamton's 'Parlor City' Nickname Explained," *Binghamton Press & Sun-Bulletin*, January 17, 1964.

16   "clean, fair, and grand": Ed Aswad and Suzanne Meredith, *Binghamton* (Images of America; Arcadia Publishing, 2001), 11.

16   In 1898: Arthur Reblitz and David Bowers, *Treasures of Mechanical Music* (Vestal Press, 1981); Reblitz, *Player Piano Servicing and Rebuilding* (Vestal Press, 1985); Reblitz and Bowers, *The Golden Age of Automatic Musical Instruments: Remarkable Music Machines and Their Stories* (Mechanical Music Press, 2001). See also Arthur Loesser, *Men, Women and Pianos: A Social History* (Simon & Schuster, 1954; repr. Dover, 1990); James R. Gaines, *The Lives of the Piano* (Holt, Rinehart and Winston, 1981); Stuart Isacoff, *A Natural History of the Piano: The Instrument, the Music, the Musicians—from Mozart to Modern Jazz and Everything in Between* (Knopf, 2011).

16   "Should one applaud?": Brian Dolan, *Inventing Entertainment: The Player Piano and the Origins of an American Musical Industry* (Rowman & Littlefield, 2009), 34.

16   **85 percent:** Reblitz, *Player Piano Servicing*, 1.

16   **popularity:** Gaines, *Lives of the Piano*; David Suisman, *Selling Sounds: The Commercial Revolution in American Music* (Harvard University Press, 2012).

16   **"every touch in technique":** Dolan, *Inventing Entertainment*, 46.

16   **one Niagara Falls company:** David Bowers, *Put Another Nickel In* (Vestal Press, 1966), 27.

17   **Jacques de Vaucanson:** Dolan, *Inventing Entertainment*, 40.

18   **alienated artists:** Dolan, 33.

18   **"menace of mechanical music":** John Philip Sousa, "The Menace of Mechanical Music," *Appleton's Magazine* 8, no. 3 (September 1906): 279.

18   **"Automatic instruments were always":** Reblitz and Bowers, *Golden Age*, 5.

18   **"Chronic mechanitis":** Dolan, *Inventing Entertainment*, 35.

18   **"endless rolls":** For technical details on the Automatic rolls see Charles H. Hamilton and George R. Thayer's 1908 US patent, "Web-controlling mechanism for self-playing instruments," US937933A.

19   **"distinctly a delight":** Reblitz and Bowers, *Treasures of Mechanical Music*, 408.

19   **High-end clubs:** See Kerry Segrave, *Jukeboxes: An American Social History* (McFarland, 2002).

19   **"most pleasant surprise":** C. Sharpe Minor to George Link, May 28, 1925.

19   **"architectural symphony":** Carl Bronson, "Artist Scores Triumph on Link Organ," February 28, 1928 (from Link Collection news clipping, unknown publication).

20   **"winged gospel":** Joseph Corn, *The Winged Gospel: America's Romance with Aviation, 1900–1950* (Oxford University Press, 1983).

20   **"air-mindedness":** Corn, *Winged Gospel*, vii, 136.

20   **Picasso:** "The Scallop Shell: 'Notre Avenir est dans l'Air,' " The Collection: Modern and Contemporary Art, Metropolitan Museum of Art.

20   **"Take possession of the air":** Wohl, *Passion for Wings*, 154.

20   **"air is an extremely dangerous":** Winston Churchill, "Adventures in the Air," *Cosmopolitan*, June 24, 1924.

20   **"Goggles, gloves":** Peter Pigott, *Brace for Impact: Air Crashes and Aviation Safety* (Dundurn, 2016), 21, 34, 38–39.

21   **internal sensations:** In one incident, a pilot realized that his aircraft was out of control and, worse, flying upside down only when his pocket watch fell out and hit his face. See Timothy P. Schultz, *The Problem with Pilots: How Physicians, Engineers, and Airpower Enthusiasts Redefined Flight* (Johns Hopkins University Press, 2018), 45.

21   **Union Pacific Railroad:** T. A. Heppenheimer, *Turbulent Skies: The History of Commercial Aviation* (John Wiley & Sons, 1995), 8.

21   **Rand McNally:** Maurer Maurer, *Aviation in the U.S. Army, 1919–1939* (United States Air Force, 1987), 36.

21   **"with a few drops of homing pigeon":** Heppenheimer, *Turbulent Skies*, 9.

21   **enemy territories:** See Conway, *Blind Landings*, 13.

21   **survey of "sky roadways":** Henry Lehrer, *Flying the Beam: Navigating the Early US Airmail Airways, 1917–1941* (Purdue University Press, 2014), 38.

22   **colorful commentary:** Bruce Etyinge and Rex Uden, *Landing Field Guide and Pilot's Log Book* (Etyinge & Uden, 1920), 53, 56.

22   **helpful hints:** Etyinge and Uden, *Landing Field Guide*, 12.

22   **"lighted airway":** Pigott, *Brace for Impact*, 38–39; Lehrer, *Flying the Beam*, 82.

23   **east to west:** Todd La Porte, "The United States Air Traffic Control System:

Increasing Reliability in the Midst of Rapid Growth" (working paper, University of California, Berkeley, 1988), 2; Heppenheimer, *Turbulent Skies*, 11.

23   "Sometimes you couldn't": Flint Whitlock and Terry L. Barnhart, *Capt. Jepp and the Little Black Book: How Barnstormer and Aviation Pioneer Elrey B. Jeppesen Made the Skies Safer for Everyone* (Savage Press, 2007), 116.

23   "He knew, for example": Whitlock and Barnhart, *Capt. Jepp*, 110.

24   400,000 people: La Porte, "Air Traffic Control System," 2.

24   "When Undertaking Very Hard Work": Lehrer, *Flying the Beam*, 82–83; Chris Forsyth and Daegan Miller, "Keep Direction by Good Methods," *Places Journal*, February 2021.

25   "Unfortunately, it wasn't": van Hoek and Link, *From Sky to Sea*, 11.

25   "just dumb enough to be a genius": Richard Whitmire, "Ed Link: I'm Not a Genius," *Binghamton Press & Sun-Bulletin*, August 15, 1976.

25   nomadic barnstormers: Paul O'Neil et al., *Barnstormers & Speed Kings* (Time-Life Books, 1981), 26.

25   "a bunch of parts": Description of Curtiss JN-4H Jenny from Smithsonian National Postal Museum, *Airmail in America*.

26   "That's a hell of a way": Lloyd Kelly and Robert Parke, *The Pilot Maker* (Grosset & Dunlap, 1970), 20.

26   Wilbur Wright trained: Rebecca Hancock Cameron, *Training to Fly: Military Flight Training, 1907–1945* (Air Force History and Museums Program, 1999), 34–37.

27   "kiwi-trainer": Max Baarspul, "A Review of Flight Simulation Techniques," *Progress in Aerospace Sciences* 27, no. 1 (1990): 1–120; Michael Moroney and William Lilienthal, "Human Factors in Simulation and Training: An Overview," in *Human Factors in Simulation and Training*, ed. D. A. Vincenzi et al. (CRC Press, 2009), 18.

28   Ruggles . . . fitness to fly: Cameron, *Training to Fly*, 265–67.

28   Ocker and Crane's conception: William C. Ocker and Carl J. Crane, *Blind Flight in Theory and Practice* (The Naylor Company, 1932).

28   two interlinked problems: Conway, *Blind Landings*, 12.

29   "And that was one of the things": Ed Link, interview by Wanda Wood, September 18, 1978, Broome County Oral History Project, Binghamton University Special Collections, 4–5.

29   Alexander Calder: Jed Perl, *Calder: The Conquest of Time; The Early Years: 1898–1940* (Knopf, 2017), 507.

Chapter 2: Roll

31   "great value as means": Ed Link, Combination Training Device for Student Aviators and Entertainment Apparatus, US patent 1,825,462, September 29, 1931.

32   "physical movements and sensations": Chihyung Jeon, "The Virtual Flier: The Link Trainer, Flight Simulation, and Pilot Identity," *Technology and Culture* 56, no. 1 (2015): 30.

33   didn't merely mimic: Jeon, "The Virtual Flier," 36.

34   Ford Model T: van Hoek and Link, *From Sky to Sea*, 48.

34   "Just a quarter": Max Hill, "Once Two-Bit Carnival Ride, Link Trainer Now Most Vital in Turning Out U.S. Fighters," *Binghamton Press*, February 27, 1943.

34   "Marion made": Ed Link, interview by Wanda Wood, 9

34   **desperate Ed Link:** van Hoek and Link, *From Sky to Sea*, 39–40.

34   **"I put a venturi":** van Hoek and Link, 43.

34   **flying billboards:** Kelly and Parke, *Pilot Maker*, 47.

35   **called themselves "airlines":** John Correll, "The Air Mail Fiasco," *Air Force Magazine*, March 1, 2008.

35   **equivalent weight of airmail:** Correll, "Air Mail Fiasco," quoting Oliver E. Allen.

35   **"crazy quilt of routes":** Conway, *Blind Landings*, 31.

35   **6,500 forced landings:** From the Smithsonian National Postal Museum exhibit, *Airmail in America*.

35   **wet their fingers:** Cameron, *Training to Fly*, 22.

35   **When pilots launched:** Even commercial flight-training schools relocated to ensure all-year flying. Cameron, *Training to Fly*, 22.

35   **two types of weather-related accidents:** Conway, *Blind Landings*, 19–20.

36   **"Aircraft that sit":** Conway, 4.

36   **"legalized murder":** Correll, "Air Mail Fiasco."

36   **Billy Mitchell:** See John Lancaster, *The Great Air Race: Glory, Tragedy, and the Dawn of American Aviation* (Liveright, 2022).

37   **Commonwealth:** van Hoek and Link, *From Sky to Sea*, 48.

37   **pivotal place:** In "Flying Training: The American Advantage in the Battle for Air Superiority against the Luftwaffe," Kenneth P. Werrell notes that the United States dramatically ramped up pilot training "at an astounding rate, from a mere 225 men who pinned on pilot wings in the last half of 1939, to 2,500 in the last quarter of 1941, rising to a peak of 29,000 in the second quarter of 1944." *Air Power History* 61, no. 1 (2014): 36.

37   **"shadows of imagination":** Samuel T. Coleridge, *Biographia Literaria; or, Biographical Sketches of My Literary Life and Opinions* (Leavitt, Lord & Company, 1834), 175.

37   **regimented real-time editing:** Lex Parrish, *Space-Flight Simulation Technology* (Howard W. Sams, 1969), 16–17.

38   **Some resisted learning:** "Link Nickelodeon Rebuild & Interview with Edwin Link Jr, November 6, 1965" (source unknown; Link Collection).

38   **"The turns did happen":** Conway, *Blind Landings*, 24.

38   **This new cognitive approach:** Schultz, *Problem with Pilots*, 58.

38   **Lacking a visible horizon:** Based on Assen "Jerry" Jordanoff, *Through the Overcast: The Art of Instrument Flying* (Funk & Wagnalls, 1938), 305–6; Jimmy Doolittle and Carroll Glines, *I Could Never Be So Lucky Again: An Autobiography* (Schiffer Publishing, 1995; originally published in 1991 by Bantam), 132–35.

39   **They sought to identify:** *Solving the Problem of Fog Flying: A Record of the Activities of the Fund's Full Flight Laboratory to Date* (Daniel Guggenheim Fund for the Promotion of Aeronautics, 1929), 5.

39   **Jimmy Doolittle:** Doolittle and Glines, *I Could Never Be*; "Jimmy Doolittle," Home of Heroes, accessed July 2023.

39   **Schneider Trophy:** Jerry Murland, *The Schneider Trophy Air Races: The Development of Flight from 1909 to the Spitfire* (Pen & Sword Books, 2021).

39   **Doolittle topped the speed record:** Richard P. Hallion, "Pioneer of Flight: Doolittle as Aviation Technologist," *Air Power History* 40, no. 4 (1993): 9.

39   **Harry Guggenheim borrowed him:** Doolittle and Glines, *I Could Never Be*, 127.

40   **five linked problems:** *Solving the Problem of Fog Flying*, 4.

40   **disperse the fog:** *Solving the Problem of Fog Flying*, 44.

40   Consolidated NY-2 biplane: See *Equipment Used in Experiments to Solve the Problem of Fog Flying: A Record of the Instruments and Experience of the Fund's Full Flight Laboratory* (Daniel Guggenheim Fund for the Promotion of Aeronautics, 1930), 6.

40   Guggenheim encouraged Doolittle: Richard Hallion, *Legacy of Flight: The Guggenheim Contribution to American Aviation* (University of Washington Press, 1977).

40   "However, despite all": Carroll Glines, *Jimmy Doolittle: Daredevil Aviator and Scientist* (Macmillian, 1972), 90.

41   Bronson Murray Cutting: Key references were related newspaper articles published on May 7, 1935, in the *New York Times, Washington Post, Baltimore Sun*, and *Chicago Daily Tribune*; Nick A. Komons, *The Cutting Air Crash: A Case Study in Early Federal Aviation Policy* (US Federal Aviation Administration, 1984); George E. Hopkins, *Flying the Line: The First Half Century of the Air Line Pilots Association* (Air Line Pilots Association, 1982).

42   "unintentional collision": Komons, *Cutting Air Crash*, 32.

43   over 40 instruments: E. J. Lovesey, "Information Overload in the Cockpit," *IEE Colloquium on Information Overload* (Institution of Electrical Engineers, 1995); R. Schroer, "Cockpit Instruments (A Century of Powered Flight: 1903–2003)," *IEEE Aerospace and Electronic Systems Magazine* 18, no. 7 (2003): 13–18.

43   "instrument explosion": E. J. Lovesey, "The Instrument Explosion—a Study of Aircraft Cockpit Instruments," *Applied Ergonomics* 8, no. 1 (1977): 23–30.

43   "The day of the throttle jockey": David Mindell, *Digital Apollo: Human and Machine in Spaceflight* (MIT Press, 2008); quoted in Schultz, *Problem with Pilots*, 220.

43   "A modern pilot": Schultz, 122.

44   To realistically reflect: Tom Cawley, "Link Trainer Secrets Are a Saga of Science but Can't Be Told Now," *Binghamton Press*, June 20, 1942.

44   flew more like an airplane: June 8, 1945, news clipping, Link Collection.

44   "Don't make the mistake": Jordanoff, *Through the Overcast*, 314–15.

44   women's opportunities in aviation: Carole Briggs, *At the Controls: Women in Aviation* (Lerner Publications, 1991).

44   Jacqueline Cochran . . . Amelia Earhart: Janene G. Leonhirth, "They Also Flew: Women Aviators in Tennessee, 1922–1950" (master's thesis, Middle Tennessee State University, 1990), 3–7.

45   "They are simply thoroughly": Amelia Earhart, *The Fun of It: Random Records of My Own Flying and of Women in Aviation* (Harcourt Brace and Company, 1932), 162.

45   *Friendship*: See Amelia Earhart, *20 Hrs. 40 Min.: Our Flight in the* Friendship (G. P. Putnam's Sons, 1928).

45   Powder Puff Derby: Gene Nora Jessen, *The Powder Puff Derby of 1929: The First All Women's Transcontinental Air Race* (Sourcebooks, 2002).

45   In one distressing instance: Corn, *Winged Gospel*, 79.

46   "Sweetie": Quoted letters from Judy Litoff and David C. Smith, "American Women in a World at War," *OAH Magazine of History* 6, no. 3 (2002): 9–10.

46   Black women: Paul Louis Dawson, "Luis De Florez and the Special Devices Division" (PhD diss., George Washington University, 2005), 204.

47   WASP: WAFS and WFTD were organized in September 1942 and incorporated into WASP in August 1943. See Sarah Byrn Rickman, *WASP of the Ferry Com-*

mand: Women Pilots, Uncommon Deeds (University of North Texas Press, 2016); Katherine Sharp Landdeck, *The Women with Silver Wings: The Inspiring True Story of the Women Airforce Service Pilots of World War II* (Crown, 2020); and related information on the websites of the Airforce Historical Support Division and National WWII Museum.

47 **"The gods must envy me":** Letter from Marion Stegeman, April 24, 1943; Litoff and Smith, 11.

47 **Link Trainer instructors:** Corn, *Winged Gospel*, 89.

47 **Eugene Jacques Bullard:** Jim Haskins, *Black Eagles: African Americans in Aviation* (Scholastic, 1995), 5–9.

48 **Bessie Coleman:** Haskins, *Black Eagles*, 17–43; Doris L. Rich, *Queen Bess: Daredevil Aviator* (Smithsonian Institution Press, 1993); Flint Whitlock, "Racial Discrimination against Pilots: An Historical Perspective," in *Ethical Issues in Aviation*, ed. E. Hoppe (Ashgate, 2011), 137–44.

48 **Hubert Fauntleroy Julian:** David Shaftel, "The Black Eagle of Harlem," *Air & Space*, December 2008.

48 **William Jenifer Powell:** Haskins, *Black Eagles*, 50–66.

49 **"While originally it"** . . . **Banning and Allen:** Julia Lauria-Blum, "James H. Banning and the Flying Hoboes Transcontinental Flight," *Metropolitan Airport News*, May 13, 2020.

49 **"with the gracefulness of a bird":** Banning article in the *Pittsburgh Courier*, December 17, 1932; reprinted in Joseph J. Corn, ed., *Into the Blue: American Writing on Aviation and Spaceflight* (Library of America, 2011), 78–81.

49 **Tuskegee:** See Charles E. Francis, *The Tuskegee Airmen: The Men Who Changed a Nation*, 5th ed., ed. Adolph Caso (Branden Books, 2008; first published 1955); Charles W. Dryden, *A-Train: Memoirs of a Tuskegee Airman* (University of Alabama Press, 1997); Samuel L. Broadnax, *Blue Skies, Black Wings: African American Pioneers of Aviation* (University of Nebraska Press, 2008).

49 **"I always heard":** Quote as reported in David Stout, "Charles Anderson Dies at 89; Trainer of Tuskegee Airmen," *New York Times*, April 17, 1996.

49 **"Keep us flying!"** . . . **"Your letters and gifts":** "The Tuskegee Airmen: Training and Stateside Experiences," website for the Tuskegee Airmen National Historical Site, National Park Service, accessed July 2023.

51 **dead reckoning:** H. H. Shufeldt and G. D. Dunlap, *Piloting and Dead Reckoning*, 4th ed., ed. Bruce Allan Bauer (Naval Institute Press, 1999).

51 **higher altitudes:** Philip Van Horn Weems, *Air Navigation* (McGraw-Hill, 1943), 355.

51 **almanac:** Boyden Sparkes, "Teaching 'Lindy' Navigation," *Popular Science Monthly*, August 1928, 53–54, 108.

51 **presenting geography:** Edwin A. Link, "Helping Hand," *Connecting Link* 4, no. 1 (January 1947).

Chapter 3: Yaw

55 **Morton and Watson:** Taken from National Transportation Safety Board's 1968 Aircraft Accident Report: Delta Air Lines, Inc. DC-8, N802E, Kenner, Louisiana, March 30, 1967 (Adopted: December 20, 1967), Washington, DC: Department of Transportation. Related syndicated news reports were consulted from the *Lub-*

bock Avalanche-Journal (March 31, 1967) and the former *Argus Fremont* (April 1, 1967).

58   **Eastern Air Lines:** From Civil Aeronautics Board's Aviation Accident Report: Eastern Air Lines Flight 304, 7.

58   **"designated mysteries":** See, for example, Adriane Quinlan, "50 Years after Eastern Air Lines Flight 304 Crashed into Lake Pontchartrain Leaving No Survivors, Something Still Remains," *Times-Picayune*, February 25, 2014.

59   **"nothing but a mechanical spy":** Pigott, *Brace for Impact*, 101.

59   **David Warren:** Based on Marcus Williamson's July 31, 2010, obituary in the *Independent* and Roger Connor's presentation "The Black Box: Creating Resiliency in Air Transport for 77 Years: But, Where Did It Come From?," 2014 IEEE/AIAA 33rd Digital Avionics Systems Conference.

60   **best place . . . "never saw an airplane":** Scott M. Fisher, "Father of the Black Box," HistoryNet, June 29, 2017.

60   **"indestructible machine":** Quotes and related background from Greg Siegel, "Technologies of Accident: Forensic Media, Crash Analysis, and the Redefinition of Progress" (PhD diss., University of North Carolina at Chapel Hill, 2005), 73, 76–80.

60   **withstand impact shocks:** Greg Siegel, *Forensic Media: Reconstructing Accidents in Accelerated Modernity* (Duke University Press, 2014), 82.

60   **"live truthfully under imaginary circumstances":** Larry Silverberg, *The Sanford Meisner Approach: An Actor's Workbook—Workbook One* (Smith and Kraus, 1994), 9.

60   **analog processors:** Masaaki Hirooka, *Innovation Dynamism and Economic Growth: A Nonlinear Perspective* (Edward Elgar Publishing, 2006), 245–46.

61   **the simulator:** John Killick, "Proxy Flight," *Pegasus*, January 1954, 8.

61   **VAMP:** Kelly and Parke, *Pilot Maker*, 136–37.

61   **"simulation syndrome":** From the title of James Der Derian's "The Simulation Syndrome: From War Games to Game Wars," *Social Text* 24 (1990): 187–92.

62   **specialized simulators:** P. W. Caro, "Flight Crew Training Technology: A Review," NASA Ames Research Center, prepared by Seville Training Systems, 1984.

62   **"In the culture of simulation":** Sherry Turkle, *Life on the Screen* (Simon & Schuster, 1995), 24.

63   **"Microsoft Flight Simulator":** For more on Bruce Artwick and the philosophy behind the product design, see Preston Lerner, "Pilot Program," *Air & Space Quarterly*, March 22, 2023.

64   **forms of fidelity:** Don Harris, *Human Performance on the Flight Deck* (Ashgate, 2011), 126–33.

64   **g-forces as experienced:** Kelly and Parke, *Pilot Maker*, 70.

64   **"lack of vertigo":** Jeon, "The Virtual Flier," 46.

65   **"The pilots become so preoccupied":** Philip Klass, "Link Simulator Boosts B-47 Potential," *Aviation Week* 56, no. 24 (June 16, 1952).

65   **"pucker factor":** Quote from Robert W. Weight, "Flight Simulators in the RAAF: A Bold Step Forward or Back to the Future?," *Wings* (Spring 2013): 36; Peter Hobbins, "Emulating the 'Pucker Factor': Faith, Fidelity and Flight Simulation in Australia, 1936–58," *Journal of Transport History* 44, no. 1 (2022): 3–26.

65   **United Airlines and Continental:** Andy Pasztor and Susan Carey, "United Continental Pilots Split on Training Simulators," *Wall Street Journal*, June 20, 2011.

66 **"Off we go into the wild blue yonder":** From the official song of the US Air Force, often simply referred to as "Wild Blue Yonder."

67 **"gamify" the experience:** Timothy Lenoir, "All but War Is Simulation: The Military-Entertainment Complex," *Configurations* 8, no. 3 (2000): 295.

67 **"Not that I wish to":** Ed Link to Margaret Weems, April 10, 1944.

68 **Alex Seiden:** National Research Council, *Modeling and Simulation: Linking Entertainment and Defense* (National Academies Press, 1997), 168–69.

68 **Mountains cannot resemble . . . operating environment:** See, for example, Patrick Crogan, *Gameplay Mode: War, Simulation, and Technoculture* (University of Minnesota Press, 2011); and Ralph Norman Haber, "Flight Simulation," *Scientific American* 255, no. 1 (1986): 96–103.

69 **their latest software:** For an in-depth review see David Allerton, *Flight Simulation Software: Design, Development and Testing* (John Wiley & Sons, 2023).

70 **"dromoscopic vision"; "The world flown over":** Quotes from Crogan, *Gameplay Mode*, 37, 41. Based on Paul Virilio, *Negative Horizon*, trans. Michael Degener (Continuum, 1989), 105–19.

71 **visual detail:** Ivan Sutherland, "The Ultimate Display," *Proceedings of the International Federation for Information Processing Congress* 65, no. 2 (1965): 506–8.

71 **"been adding one thing":** From Ed Link's letter to C. S. Jones, October 25, 1941.

72 **"must abandon the notion":** From Eduardo Salas, Clint A Bowers, and Lori Rhodenizer, "It Is Not How Much You Have but How You Use It: Toward a Rational Use of Simulation to Support Aviation Training," *International Journal of Aviation Psychology* 8, no. 3 (1998): 197–208.

72 **"I sort of lost interest":** From Mary Beth Herzog, *Vero Beach Press Journal*, May 1, 1977.

73 **"I started as a grease monkey":** From Clarence E. Lovejoy, *New York Times*, February 23, 1956.

Chapter 4: Surge

75 **White House Cabinet Room:** From John F. Kennedy Library, President's Office Files, Presidential Recordings Collection, Tape 63A, November 21, 1962, from Presidential Recordings, Miller Center, University of Virginia. I encountered this in Erinn Catherine McComb's dissertation, "Why Can't a Woman Fly? NASA and the Cult of Masculinity, 1958–1972" (Mississippi State University, 2012), 274.

75 **political symbolism:** See Walter A. McDougall, *The Heavens and the Earth: A Political History of the Space Age* (Basic, 1985); Teasel Muir-Harmony, *Operation Moonglow: A Political History of Project Apollo* (Basic, 2020).

78 **A crucial hurdle:** See John McLeod, "Manned Spacecraft Simulation," *Proceedings of the May 21–23, 1963, Spring Joint Computer Conference* (Detroit, Michigan, 1963), 401–9.

78 **ENIAC; "spurring rumors":** Wikipedia, "ENIAC," August 2023; Gregory Farrington, "ENIAC: Birth of the Information Age," *Popular Science*, March 1996, 74.

78 **$2 million each:** The computers were IBM 9600/9604 family. Mark-I and Mark-II could barely compute the equations of motions at the requisite speed.

78 **instructions per second:** "Sampling frequency" specifies performing at least twice as quickly as the natural frequency of the simulated process. Abbey's team needed to execute the equations 10 times faster to produce real-time simula-

tions of the variables. See Link Capabilities Statement, April 1963, courtesy of TechWorks!

79 **systems integration:** Manfred von Ehrenfried, *Apollo Mission Control: The Making of a National Historic Landmark* (Springer International Publishing, 2018), 130.

79 **human impulse:** Based on Stanley H. Goldstein, "Astronaut Training: An Administrative History of Projects Mercury, Gemini and Apollo" (master's thesis, University of Colorado, 1984), 3–4; and Goldstein's *Reaching for the Stars: The Story of Astronaut Training and the Lunar Landing* (Praeger, 1987).

80 **time in simulators:** See C. H. Woodling et al., "Apollo Experience Report: Simulation of Manned Space Flight for Crew Training" (NASA Technical Note D-7112, 1973), 1–3.

80 **"The final simulation":** Gene Kranz, *Failure Is Not an Option: Mission Control from Mercury to Apollo 13 and Beyond* (Simon & Schuster, 2000).

80 **nearly 700 switches:** Hamish Lindsay, *Tracking Apollo to the Moon* (Springer-Verlag London, 2001), 206–7.

80 **sophisticated visual experience:** Based on Paul Ceruzzi, *Beyond the Limits: Flight Enters the Computer Age* (MIT Press, 1989), 170; James Tomayko, *Computers Take Flight: A History of NASA's Pioneering Digital Fly-by-Wire Project* (NASA History Office, 2000); Joe Dahm, *Evening Press* (Binghamton), November 24, 1964, 17, 32.

81 **10 tons devoted:** Kelly and Parke, *Pilot Maker*, 160.

81 **"Apollo gave new":** Chiabotti, "The Glorified Link," 61, 63.

81 **14-hour days:** See David M. Harland's "Preparations," in *The First Men on the Moon: The Story of Apollo 11* (Springer Praxis Books, 2007).

82 **When the *Eagle* landed:** *Apollo 11* Technical Air-to-Ground Voice Transmission (GOSS NET 1), Tape 70, no. 24, p. 377.

82 **"normal design" . . . "radical design":** Walter Vincenti, "Engineering Knowledge, Type of Design, and Level of Hierarchy: Further Thoughts About What Engineers Know . . . ," in *Technological Development and Science in the Industrial Age*, ed. P. Kroes and M. Bakker (Kluwer Academic Publishers, 1992), 17–34.

83 **weight-loss project:** Courtney Brooks, James M. Grimwood, and Loyd S. Swenson, *Chariots for Apollo: The NASA History of Manned Lunar Spacecraft to 1969* (NASA, 1979), 172–75.

84 **"organization, education, and training":** Jean-Jacques Servan-Schreiber, quoted in Stephen Johnson, *The Secret of Apollo: Systems Management in American and European Space Programs* (Johns Hopkins University Press, 2002), 5.

84 **success of such projects:** Stephen Johnson, "Philosophical Observations and Applications in Systems and Aerospace Engineering," in *Engineering and Philosophy: Reimagining Technology and Social Progress*, ed. Z. Pirtle, D. Tomblin, and G. Madhavan (Springer, 2021), 94.

84 **"We can lick gravity":** Often attributed to Wernher Von Braun.

84 **"Spacecraft that":** Johnson, *Secret of Apollo*, 4.

84 **six-step procedure:** Brooks, Grimwood, and Swenson, *Chariots for Apollo*, 169.

85 **"insurance for technical success":** Johnson, *Secret of Apollo*, 223.

85 **"Military officers":** Johnson, 17.

85 **"We were lucky all the time":** Vincent Davis, "NASA Retiree Reflects on Apollo 13 Crisis," *San Antonio Express-News*, April 10, 2010.

85   **"no training rhythm to it"**: Francis E. "Frank" Hughes, interview by Rebecca Wright, Houston, Texas, September 10, 2013, Edited Oral History Transcript, NASA Johnson Space Center Oral History Project.

86   **"Houston, we've had a problem"**: *Apollo 13* Technical Air-to-Ground Voice Transcription, NASA Apollo Spacecraft Program Office, Manned Spacecraft Center, Houston, Texas, April 1970.

85   **The Link simulator**: Based on "Deliverance from Disaster," *Connecting Link* (Summer 1970): 2–5.

87   **Apollo 13 needed**: See Jim Lovell and Jeffrey Kluger, *Apollo 13* (Mariner Books, 2006); Fred Haise and Bill Moore, *Never Panic Early: An Apollo 13 Astronaut's Journey* (Smithsonian Books, 2022).

87   **"successful failure"; "unsung heroes"**: Frank Hughes's foreword for Brian Woycechowsky, *Lunar Module Moon-Referenced Equations of Motion* (Center for Technology & Innovation, 2021), vii.

88   **"Makes you feel kind of creepy"**: Kurt Vonnegut, *Player Piano* (1952; Dial Press, 1999), 32.

88   **operational preservation**: John Chilvers, "Curatorial Lessons from Other Operational Preservationists: Towards a Methodology for Computer Conservation," *Proceedings of Making IT Work* (British Computer Society, National Museum of Computing, 2017), 44–53.

89   **Whirlpool**: See Gerald Smith, "How a Household Name Got Its Start in Binghamton," *Binghamton Press & Sun-Bulletin*, June 16, 2016.

90   **military flight simulators**: Dave Peters (cofounder of Diamond Visionics) and Richard Mecklenborg also significantly contributed to simulation through the wide-angled collimated window display technology, with Mecklenborg also advancing the dome display for the British vertical take-off aircraft.

91   **Bennett wrote a monograph**: *Visualizing Software: A Graphical Notation for Analysis, Design, and Discussion* (Marcel Dekker, 1992), 3–6.

92   **"Another flaw"**: Kurt Vonnegut, *Hocus Pocus* (1990; Berkley, 1997), 238.

92   **"Any sufficiently"**: Debbie Chachra's June 14, 2017, tweet.

Chapter 5: Sway

95   **"part fish and part fowl"**: Coles Phinizy, "The Missing Link," *Sports Illustrated*, July 15, 1957, 27–29.

95   **Linkanoe**: Kelly and Parke, *Pilot Maker*, 87–88.

96   **"water and air"**: Whitmire, "Ed Link: I'm not a Genius."

96   **Looe Key**: van Hoek and Link, *From Sky to Sea*, 84–86.

96   **"I took up golf"**: Phinizy, "The Missing Link."

96   **"A beautiful sea garden"**: van Hoek and Link, *From Sky to Sea*, 87.

97   **"When we made those"**: Marion Clayton Link, *Sea Diver: A Quest for History under the Sea* (University of Miami Press, 1964), 3.

97   **"I was able to swim"**: Marion Clayton Link, *Sea Diver*, 26.

97   **"I found out"**: Quote from Kelly and Parke, *Pilot Maker*, 90.

98   **Columbus's first landfall**: Ed Link, "My Flights over Columbus' Routes, Dictated aboard 'Sea Diver,'" transcribed April 20, 1955, 1, 6–7; Link, "Big Argument about His First Landing," *Life*, October 12, 1959; Florida Institute of Technology Link Special Collection, and based on Ed and Marion Link, *A New Theory on Columbus' Voyage through the Bahamas* (Smithsonian Institution, 1958).

98   "sailor's imagination": "Columbus Landing Spot Disputed," *New York Times*, March 10, 1958, 1, 31.

98   "a place of spectacular": van Hoek and Link, *From Sky to Sea*, 139.

98   "It took guts": van Hoek and Link, 148.

99   British cannon: van Hoek and Link, 159.

99   World Heritage Site: UNESCO, with acknowledgment to Ed Link, noted there's "no national or regional comparison for Port Royal as it is the only authentic sunken city in the Western Hemisphere."

99   *Sea Diver II*: Marion Clayton Link, *Windows in the Sea* (Smithsonian Institution Press, 1973).

100   explore Caesarea: Charles T. Fritsch and Immanuel Ben-Dor, "The Link Expedition to Israel, 1960," *Biblical Archaeologist* 24, no. 2 (1961): 50–59.

100   Apollonius of Tyana: Charles T. Fritsch with Glanville Downey et al., eds., *Studies in the History of Caesarea Maritima* (Scholars Press, 1975), 7.

100   engineering spectacle: Jerry Handte, "Links Hope to Find Ancient Religious Tablets on Expedition to Site of Caesarea," *Sunday Press* (Binghamton), February 21, 1960.

100   Book of Jonah: From Jonah 1:5 (New King James Version), quoted in Fritsch and Ben-Dor, "Link Expedition to Israel," 53.

100   size of specific structures: Ed Link's undated manuscript (likely circa 1960), "Survey Trip to Israel," Florida Institute of Technology Link Special Collection, 2.

101   relics to the Israeli government: Eliav Simon, "Scientific Ship Searches Israeli Harbor for Lost City Herod Built," *Milwaukee Journal*, July 26, 1960.

101   underwater living: Based on James Goff, "Aerospace Concepts Applied to Deep Submergence Vehicle Simulation," *SAE Transactions* 76 (1968): 1239–43.

101   Bushnell: Brenda Milkofskym, "David Bushnell and His Revolutionary Submarine," ConnecticutHistory.org, September 6, 2019.

103   Jacques Cousteau: See Jacques Cousteau and Frederic Duma, *The Silent World* (Harper, 1953); Brad Matesen, *Jacques Cousteau: The Sea King* (Pantheon, 2009).

103   "big meal of cabbage": Quote from Peter Fairley, "Caroline, The Goat, Paves the Way," *London Evening Standard*, October 20, 1962.

103   "Man-in-Sea": A key reference was Ed Link, "Our Man-in-Sea Project," *National Geographic*, May 1963.

103   prospectus to the US Navy: Based on the Man-in-Sea program prospectus from General Precision, Inc., February 14, 1963, and as described by Andrew Clark, "The Link Legacy," *Marine Technology Society Journal* 49, no. 6 (2015): 43–44.

104   "We cannot afford to neglect": From Ed Link, draft untitled report on deep submergence, September 6, 1963, Florida Institute of Technology Link Special Collection, 6.

104   USS *Thresher*: See F. N. Spiess and A. E. Maxwell, "Search for the 'Thresher,'" *Science* 145, no. 3630 (1964): 349–55.

104   "should not be viewed" . . . "good engineering": Francis Duncan, *Rickover and the Nuclear Navy: The Discipline of Technology* (Naval Institute Press, 1990), 85.

105   "it could be possible": Link, draft untitled report, 6.

105   "We should not have lost": Joe MacInnis, *Breathing Underwater: The Quest to Live in the Sea* (Viking Canada, 2004), 17.

106   the SPID: Based on articles in the *Miami News*, *San Francisco Chronicle*, and *Binghamton Press* from early July 1964.

106   "I have always": Quoted in MacInnis, *Breathing Underwater*, 74; Robert Sténuit,

The Deepest Days: A Remarkable Odyssey of Undersea Adventure and the Longest, Deepest Dive Ever Made (Hodder and Stoughton, 1966).

106 **"Our mission was accomplished":** Ed Link, "Outpost under the Ocean," *National Geographic*, April 1965, 532.

107 **maximum maneuverability:** Goff, "Aerospace Concepts Applied."

107 **"Living in the sea":** Wallace Mitchell, "Leave Living in the Sea to Fish, Says Scientist," *Honolulu Advertiser*, July 13, 1967.

107 **"Volkswagens of the deep":** Ed Link, "Working Deep in the Sea," in *The World in 1984*, ed. N. Calder (Penguin Books, 1964), 1:103–5.

108 **Great Stirrup Cay:** Tom Huser, *Palm Beach Post-Times*, March 10, 1968.

108 **highly maneuverable submersible:** See R. Frank Busby, *Manned Submersibles* (Office of the Oceanographer of the Navy, 1976).

109 *Johnson-Sea-Link*: Based on Timothy Askew, "Submersibles for Science: Johnson-Sea-Link I & II," in *Proceedings of Oceans 1984* (Marine Technology Society and IEEE Ocean Engineering Society, Washington, DC, September 1984), 612–16.

109 **June 18, 1973:** June 19–20, 1973, articles in the *Evening Press* and *Sun-Bulletin* (Binghamton), and in the *Floridian* (by Don North), January 11, 1976; and Clark, "The Link Legacy."

110 **"At this moment":** MacInnis, *Breathing Underwater*, 219.

110 **"this loss need never":** van Hoek and Link, *From Sky to Sea*, 306.

110 **"work blind":** Tom Cawley, *Evening Press* (Binghamton), April 3, 1975.

110 **"It made us realize":** Mary Beth Herzog, *Vero Beach Press Journal*, May 1, 1977.

110 **"I'm demanding":** Annie Laurie Morgan, unpublished draft profile of Marilyn Link, 1977.

111 **"Our seas are":** Tom Cawley, "Edwin Link Starts New Career at 71," *Binghamton Press*, 1975 [undated in author's copy].

111 **"haven't taken"; "There is ignorance at all levels":** Barry Holtzclaw, *Sun-Bulletin* (Binghamton), July 26, 1971.

112 *Johnson-Sea-Link*s **helped locate:** Timothy Askew's 1987 conference paper, "Johnson-Sea-Link Submersibles' Role in the Challenger Recovery," *Proceedings of Oceans '87* (Marine Technology Society and IEEE Ocean Engineering Society, Nova Scotia), 1225–28.

112 **"Were it not for CORD":** Clark, "The Link Legacy," 52.

112 **"In the offshore business":** Clark, 50; General Precision, Inc., "The *Man in Sea* Program," prospectus prepared for US Navy, February 14, 1963.

113 **"Few major fields":** Clark, 45.

113 *Titanic*: See Joe MacInnis, *Titanic in a New Light* (Thomasson-Grant, 1992).

Chapter 6: Heave

115 **"probably received":** Moroney and Lilienthal, "Human Factors in Simulation and Training," 4.

115 **early simulators were games:** Based on Pat Harrigan and Matthew Kirschenbaum, eds., *Zones of Control: Perspectives on Wargaming* (MIT Press, 2016); Charles Homans, "War Games: A Short History," *Foreign Policy*, August 31, 2011.

116 **"approximation of war"; "was always just an approximation":** Jon Peterson, "A Game out of All Proportions: How a Hobby Miniaturized War," in Harrigan and Kirschenbaum, *Zones of Control*, 4, 23.

116 **Dungeons & Dragons:** See Jon Peterson, *Game Wizards: The Epic Battle for*

Dungeons & Dragons (MIT Press, 2021); Ben Rigg, *Slaying the Dragon: A Secret History of Dungeons & Dragons* (St. Martin's Press, 2022).

116 **gaming the unthinkable:** See Sharon Ghamari-Tabrizi, "Simulating the Unthinkable: Gaming Future War in the 1950s and 1960s," *Social Studies of Science* 30, no. 2 (2000): 163–223.

116 **Jack Thorpe:** Jack Thorpe, "Future Views: Aircrew Training 1980–2000," Bolling Air Force Base, Washington, DC, September 15, 1987. See Thorpe's related retrospective, "Trends in Modeling, Simulation, & Gaming: Personal Observations about the Past Thirty Years and Speculation About the Next Ten," paper presented at the 2010 Interservice/Industry Training, Simulation, and Education Conference (I/ITSEC).

117 **high contextual complexity:** Lenoir, "All but War Is Simulation," 311–12.

117 **"Alice's looking glass":** Thorpe, "Future Views," 5.

117 **"Instead of communicating":** Thorpe, 8–9.

118 **"Fast, approximate, and cheap":** Thorpe, "Trends in Modeling," 33.

118 **Pentagon and Hollywood:** Based on Sharon Ghamari-Tabrizi, "The Convergence of the Pentagon and Hollywood," in *Memory Bytes: History, Technology, and Digital Culture*, ed. L. Rabinovitz and A. Geil (Duke University Press, 2004).

118 **"flying carpet":** From Project ODIN (1990–91), described in Thorpe, "Trends in Modeling," 11. Related reading included Duncan Miller and Jack Thorpe, "SIMNET: The Advent of Simulator Networking," *Proceedings of the IEEE* 83, no. 8:1114–23; see also Crogan, *Gameplay Mode*, and Sharon Weinberger, *The Imagineers of War* (Knopf, 2017).

118 **"distant intimacy":** John Williams, "Distant Intimacy: Space, Drones, and Just War," *Ethics & International Affairs* 29, no. 1 (2015): 93–110.

119 **"reorientation is paired with":** Crogan, *Gameplay Mode*, 57.

119 **"War is the province of uncertainty":** Quoted in John Rhea, "Planet Simnet," *Air Force*, August 1989, 64.

119 **"What's the difference between fighting":** Quote by Richard Lindheim, executive vice president of Paramount Digital Entertainment, from Ghamari-Tabrizi, "Convergence of the Pentagon and Hollywood," 165.

119 **"light as a feather":** Walt Whitman, *Democratic Vistas and Other Papers* (Walter Scott, 1888), 71.

120 **hourly training costs:** The US Air Force estimate is Dehmel-type simulator for Boeing's B-50 Superfortress, from Hobbins, "Emulating the 'Pucker Factor.'"

120 **organ factory:** Barry Holtzclaw, *Sun-Bulletin* (Binghamton), July 26, 1971; "The Story of the Link Orchestral Organ," Roberson Center for the Arts and Sciences, possibly 1968; Billy Nale, "The Renaissance of the Organ," *Music Magazine* (late 1960s [undated in author's copy]).

120 **"I've had a fun career":** Tom Cawley, *Binghamton Press*, September 30, 1964.

121 **"Don't you ever believe":** "Innovators," *Business in New York State*, November/December 1968, 6–8.

121 **Telford:** Samuel Smiles, *Lives of the Engineers: History of Roads* (John Murray, 1874), 206.

121 **EJ:** William Inglis, *George F. Johnson and His Industrial Democracy* (Huntington, 1935); Kristina Wilcox, "Factory Town in Transition: A Community's Reaction to Change" (PhD diss., Georgetown University, 2015).

121 **"nifty shoes":** Ed Aswad and Suzanne Meredith, *Endicott-Johnson* (Images of America; Arcadia Publishing, 2003), 14.

122 **"except ordinary decency"**: Richard Sherwood Saul, "An American Entrepreneur: George F. Johnson" (PhD diss., Syracuse University, 1966), 295.

122 **"negotiated loyalty"**: Gerald Zahavi, "Negotiated Loyalty: Welfare Capitalism and the Shoeworkers of Endicott Johnson, 1920–1940," *Journal of American History* 70, no. 3 (1983): 602–20.

122 **"efficiency expenses"; "veritable beehive of industry"**: Zahavi, *Workers, Managers, and Welfare Capitalism: The Shoeworkers and Tanners of Endicott Johnson, 1890–1950* (University of Illinois Press, 1988), 38, 6.

122 **"EJ Homes"; "There can be no security"**: Mark Simonson, "Shoe Firm Put Workers' Housing First in Endicott," *Daily Star* (Oneonta), May 31, 2008.

123 **"godfather of sports"**: Zahavi, *Workers, Managers, and Welfare Capitalism*, 51.

123 **"workingman's advocate" and "the working people"**: Zahavi, 51.

123 **"My father's family"**: Wilcox, "Factory Town in Transition," 34.

124 **Thomas J. Watson Sr.**: Thomas Graham Belden and Marva Robins Belden, *The Lengthening Shadow: The Life of Thomas J. Watson* (Little, Brown, 1962), 157, 166, 167, 194.

124 **"No errors"**: McGuire and Osterud, *Working Lives*, 81. A similar commitment to accuracy proved decisive in the 1890 census, conducted in just six months with engineer Herman Hollerith's punch-card tabulator, with ancestral roots in the Jacquard loom and player pianos.

124 **tabulating machines**: Belden and Belden, *Lengthening Shadow*, 162–63; related useful background in Lars Heide, *Punched-Card Systems and the Early Information Explosion, 1880–1945* (Johns Hopkins University Press, 2009).

124 **With the motto THINK**: Belden and Belden, *Lengthening Shadow*, 158.

125 **"ever onward IBM"**: Belden and Belden, 135–36.

125 **religious tones**: Belden and Belden, 127.

125 **IBM newspaper mused**: Belden and Belden, 128–29. Quote from Elbert Green Hubbard, *National Poultry, Butter & Egg Bulletin*, October 1923, 13.

125 **"He always shaved" ... "people caring about"**: Quotations in this paragraph from Belden and Belden.

126 **"We don't have to simulate"**: Link Aviation Devices, Inc., Inter-Departmental Correspondence, Subject: "The Future of the Company," November 8, 1944, 1, Link Collection.

126 **"Many English workers"**: *Binghamton Press* article, May 1949 [undated in author's copy].

127 **Ronald Reagan**: From September 12, 1984, remarks at Reagan-Bush rally in Endicott, New York, Ty Cobb Field at Union-Endicott High School, Ronald Reagan Presidential Library and Museum.

127 **2012 well-being survey**: Gallup-Healthways Well-Being Index; Binghamton followed McAllen-Edinburg-Mission, Texas. See related reporting by Angela Haupt in *US News & World Report*, March 13, 2012.

127 **"They were confident"**: Wilcox, "Factory Town in Transition," 8.

127 **1,300 workers**: Online IBM archives.

128 **"biggest eyesores"**: Jeff Platsky, "Wanted: Developers for IBM Country Club, Assistance Provided," *Binghamton Press & Sun-Bulletin*, June 20, 2019.

128 **"Remember: the past won't"**: Joseph Brodsky, "San Pietro," in *A Part of Speech*, trans. Barry Rubin (Farrar, Straus and Giroux, 1980), 147.

128 **"We kind of grew up"**: Katie Sullivan Borrelli, "Endicott Johnson Shoe Co.'s

Former Employees Say Company Was 'Like Family,' " *Binghamton Press & Sun-Bulletin*, October 16, 2018.

130 **"walked up and down"**: Ronald Capalaces, *When All the Men Were Gone: World War II and the Home Front; One Boy's Journey through the War Years* (Lazarus Publishing, 2010), 25.

130 **"the past jostling the present"**: Quote from Julia Van Haaften, *Berenice Abbott: A Life in Photography* (W. W. Norton, 2018), 196.

132 aerodynamic Art Deco bus station: The Streamline Moderne style, the website Treasures of the Tier notes, was "intended to depict aerodynamics and a sense of speed, the design is attributed to Louisville architect William S. Arrasmith, who designed over sixty moderne Greyhound terminals in his career, of which only a half-dozen exist today." See Frank Wrenick and Elaine V. Wrenick, *The Streamline Era Greyhound Terminals: The Architecture of W. S. Arrasmith* (McFarland, 2007).

132 **Armando Dellasanta**: Little Venice's Manhattan and Binghamton Rooms feature the impressionist's energetic paintings of the cities that gave him the names "Binghamton's Van Gogh" and "urban Monet."

Epilogue: Thinking Inside the Box

135 **"The many fictional inner journeys"**: "Dire Cartographies: The Roads to Ustopia," in *In Other Worlds: SF and the Human Imagination* (Doubleday, 2011), 70.

136 **"depth technology"**: Judith Roof, "Depth Technologies," in *Technospaces: Inside the New Media*, ed. Sally R. Munt (Continuum, 2001), 21–22.

136 **"It exists for only"**: Michael Kelly, "David Gergen, Master of the Game," *New York Times Magazine*, October 31, 1993.

136 **"Great events"**: Quote from William Gass, "Paul Valéry: The Later Poems and Prose," *New York Times*, August 27, 1972.

136 exalted status: Corn, *Winged Gospel*, 26.

137 **"Due recognition"**: Charles Taylor, "The Politics of Recognition," in *Multiculturalism: Examining the Politics of Recognition*, ed. A. Gutmann (Princeton University Press, 1994), 25–26.

137 fly without motors: Wilbur Wright's May 13, 1900, letter to Octave Chanute, Library of Congress Manuscript Division, Octave Chanute Papers.

138 **"While he did not go"**: "Edwin A. Link, Prepared by Ralph E. Flexman," April 5, 1985, unpublished.

138 **"By missing a university education"**: MacInnis, *Breathing Underwater*, 20.

138 **"Ed Link is the most"**: Phinizy, "The Missing Link," 27–29.

138 **"No-o-o, no, I'm not"**: Whitmire, "Ed Link: I'm Not a Genius."

138 **"Define the problem"** . . . **"can result in"**: 1979 interview in the *Sun-Bulletin* (Binghamton), September 8, 1981.

140 **Kālidāsa**: The lyrical Sanskrit gem "Meghaduta," or "cloud messenger," from the fourth to fifth century CE, has been translated in Thomas Clark's 1882 book *Meghaduta: The Cloud Messenger* (Trubner and Co.). See also Arthur W. Ryder's *Kalidasa: Translations of Shakunthala and Other Works* (J. M. Dent & Sons/E. P. Dutton).

Index

99th Pursuit Squadron, 49–50

Abbey, Gerald Gene, 75–77, 78–79, 90–91
abstraction, 91
acrobatic showmen, 25–26, 28–29, 33–34, 48
aerial advertising, 34
aeroneurosis, 59
Air Mail Act, 34–35
Air Navigation (Weems), 51
air shows, 25–26, 33–34, 48. *See also* barnstormers
airmail pilots
    Air Mail Act, 34–35
    air mapping, 21–22
    Army Air Corps, 36
    lighted airways, 22–23
    Lindbergh as, 11, 36
    navigation routes, 35
    road maps, 21
    standardized directions and, 23–24
Airport (film), 61
Aldrin, Buzz, 82
Allen, Thomas, 49
Aloha, 13, 14, 51
Anderson, Charles Alfred, 49
Antoinette Flying School, 27
Apollo 1, 79
Apollo 11, 81–82, 83–84
Apollo 13, 86–87
Apollo configuration management, 84–85
Apollo simulator program, 77, 79, 80–81, 85–86, 87, 89
aquanauts, 103
Armstrong, Neil Alden, 79, 82
Army Air Corps, 36
Automatic Musical Company, 18
automatic surveillance, 118
autopilot, 43–44
aviation
    acrobatic showmen, 25–26, 28–29, 33–34, 48
    airplanes as symbol, 20
    autopilot, 43–44

beacons, 22–23, 24, 42
Black people, 46–50
blind flight, 28, 37–39, 40–41
celestial sightings, 51–52
checklist duties, 41
contact flying, 6, 21–22, 35–36
Delta Flight 9877, 55–58
Eastern Air Lines Flight 304, 58
as entertainment, 31, 33–34, 67–68
flight-data recorders, 59–60
flying in bad weather, 6, 14–15, 35–36, 39–42, 52
gliders, 39
ground training, 7, 26–28
health of pilots, 59
instruments, 43–44
motion trainers, 27–28, 64–65
navigation skills, 51
No. 1 British Flying Training Schools, 1–4
sextants, 50
speeds per hour, 3
vertigo, 64–65
women, 44–47, 48
    See also airmail pilots; flight contests; flight simulation; flight training; Link Trainers

B-52 Stratofortress bomber, 120
balance, 38–39
Banning, James, 49
Barbarite, Gabby, 111–12
barnstormers, 25–26, 28–29, 33–34, 48
bathyscaphe, 102, 104
bathysphere, 101
beacons, 22–23, 24, 42
Bennett, Bill, 90–92
Bessie Coleman Aero Club, 49
Billing, Eardley, 27
Bingham, William, 15
Binghamton, NY, 15–16, 19, 89, 121–25, 126–33, 135–36, 139–40
Bird of Paradise, 11
black box recorders, 59–60

Black people, 3, 46–50
blind flight, 28, 37–39, 40–41
Blind Flight in Theory and Practice (Ocker and Crane), 28
Boeing Stratotanker, 80
Britain, 1–3
British cadets, 1–4
British Flying Training Schools, 1–4
Bullard, Eugene Jacques, 47–48
Bushnell, David and Ezra, 101

Cabled Observation and Rescue Device (CORD), 110, 112
Caesarea, Israel, 100–101
Calder, Alexander, 29
Capalaces, Ronald, 130
Cardullo, Frank, 32
carousel, 131
Celestial Navigation Trainer, 50, 51–52, 54, 67–68
celestial sightings, 51–52
Center for Technology & Innovation, 88
Challenger space shuttle, 112
Chaplin, Sydney, 25–26
chaturanga, 115
checklist duties, 41
Chenango Canal, 16
Church, Ellen, 45
Churchill, Winston, 20
circuses, 34
Civil Aeronautics Act, 42–43
Civil Aeronautics Authority, 43
Clark, Andrew, 112–13
Clausewitz, Carl von, 118–19
Clayton, Marion, 33, 34
Cochran, Jacqueline, 44–45
cognitive fidelity, 64
Coleman, Bessie, 48
Collins, Michael, 82
Columbus, Christopher, 97–98
computers, 60–61, 72, 78, 88
Connor, Roger, 32, 33
Consolidated NY-2 biplane, 40
contact flying, 6, 21–22, 35–36
Continental Airlines, 65
Conway, Erik, 6, 35–36, 38
Coolidge, Calvin, 4
Copeland, Royal, 42
Corn, Joseph, 19, 20, 45–46
Cousteau, Jacques, 103
cows, 21
Crane, Carl, 28, 38–39

Curtiss, Glenn, 26–27
Curtiss Jennys, 25, 28
Cutting, Bronson Murray, 41–42

Dallas Spirit, 13, 14
Davis, William, 13–14
Debain, Alexandre-François, 17
decompression chambers, 102–3, 105–6, *114*
Deep Diver, 107
Deep Submergence Systems Review Group, 105
The Deepest Days (Sténuit), 106
Deepsea Challenger, 113
Defense Advanced Research Projects Agency (DARPA), 116–17
Delta Flight 9877, 55–58
Diamond Visionics, 69–71
diving, 96–99
Dole Derby, 51
Doolittle, Jimmy, 7, 39–41
Doran, Mildred, 13, 14
Douglas DC-2 aircraft, 41–42
Douglas DC-8 aircraft, 58
Drebbel, Cornelis, 101
Dungeons & Dragons (D&D), 116

Eagle, 82
Earhart, Amelia, 45, 51
From the Earth to the Moon (Verne), 79–80
Eastern Air Lines Flight 304, 58
Edwin A. Link Field, 135
El Encanto, 11, 12
Endicott-Johnson Shoes (EJ), 121–24, 127–28, 130
engineering, 82–83. *See also* systems engineering/thinking
ENIAC, 78
Erwin, Bill, 13, 14
ethics, 118, 119
Eytinge, Bruce, 22

femininity, 44–45
fires, 79
FIT Aviation, 62–64, 66–68, 72
Flexman, Ralph, 138
flight contests, 11–14, 39, 45, 51
flight simulation
    overview, 60–66
    Abbey and, 77
    for combat, 116–17
    as cost-effective, 119–20

details of Link's, 60–61
Diamand Visionics, 69–71
FIT Aviation, 62–64, 66–68, 72
gamified, 67–69
laparoscopies, 69
level of detail, 71
VAMP, 61
See also Apollo simulator program;
    Link Trainers; low-tech simula-
    tion; space flight simulation
flight training
    Black Americans rejections, 48
    British Air Force celestial, 52
    Celestial Navigation Trainer, 50, 51–52,
        54, 67–68
    Delta Flight 9877, 55–58
    ground-based, 9, 26–28
    history of, 27–28
    Link's flying school, 31–32
    Link's training, 25–26
    motion training, 27–28, 64–65
    vs. simulation, 72
    Wright training, 26–27
    See also Apollo simulator program;
        flight simulation; Link Trainers;
        low-tech simulation; Mercury
        simulator
flight-data recorders, 59–60
Florida Atlantic University, 112–13
Flying Hoboes, 49
fog, 39–40, 41–42
Ford Motor Company, 122
France, 84
Freedom 7, 77
French Air Service, 28
Friendship, 45
Fulton, Robert, 101
"Future Views" (Thorpe), 116–17

games, 115–16
GAT-1, 92–93
Gawron, Valerie, 51–52
Gdovin, David, 69–71
gliders, 39
Goddard, Norman, 12
Goebel, Art, 13–14, 15
Golden Eagle, 13, 14, 51
Goldstein, Stanley, 79
Gordon, Louis, 45
grasscutter, 28
Greater Binghamton Airport, 135–36,
    139–40

ground-based training, 9, 26–28. *See also*
    Link Trainers
Grumman, 83
Guggenheim, Daniel and Harry, 39–40
Gulf War, 118–19

Haise, Fred, 86–87
Harbor Branch Oceanographic Institute,
    111–12, 113
Hawkins, Kenneth, 12
health, 59
Hodgson, Marion Stegeman, 47
Hoover, Herbert, 35
housing, 122–23
Hughes, Frank, 85–86
Hunt, John, 78

IBM, 123–26, 127–28
innovation, 84, 92
instruments, 43–44. *See also* blind flight
Irving, Livingston "Lone Eagle," 13

Jacquard, Joseph Marie, 17
James, Jim, 91
Japan, 37
Jefferson, Thomas, 101
Jensen, Martin, 13, 14, 15
Jeppesen, Elrey Borge, 23–24
Johnson, George F., 121–24, 131
Johnson, Seward, Sr., 109
Johnson, Stephen, 84, 85
Johnson-Sea-Links, 107–9, 112
Jones, Charles "Casey," 5, 37
Jordanoff, Assen, 44
Josephus, Flavius, 100–101
Julian, Hubert Fauntleroy, 48
Jurassic Park (film), 68

Kelly, Tom, 126, 128–29, 131
Kelsey, Benjamin, 40
Kennedy, John F., 75–76
Killebrew, Tom, 2–3
Kilmer, Sylvester Andral, 131
kiwi-trainer, 27
Kranz, Gene, 80
kriegsspiele, 115
Ku Klux Klan, 122

Landing Field Guide and Pilot's Log Book
    (Eytinge), 22
laparoscopies, 69
Lash, John, 89–90

licensing, 45
lighted airways, 22–23, 42
Lilienthal, Michael, 115
Lindbergh, Charles "Slim," 15, 29, 36, 50–51
Lindbergh, Jon, 106
Link, Clayton, 103, 109
Link, Edwin A., Jr.
  overview, 136–38
  about, 8, 24–25
  aerial advertising, 34
  air shows, 33–34
  airplane-repair shop, 33
  aquatic device, *134*
  Clark on, 113
  commemoration, 129, 133, *142–43*
  computers and, 72
  CORDs, 110, 112
  death of, 121
  death of Clayton, 109–10
  decompression chambers, 102–3, 105–
    6, *114*
  as diver, 96–101, 103, 108
  as employer, 126
  as fisherman, *95–96*
  Flexman on, 138
  flying into Newark, 5–6
  flying lessons, 25–26, 28–29
  flying school of, 31–32
  government regulators and, 71
  internships at Harbor Branch, 113
  MacInnis on, 113
  Marion and, 33
  McKee on, 138
  ocean engineering at Florida Atlantic
    University, 112–13
  on oil industry, 111
  organ collecting, 120
  philosophy of systems engineering, 90
  Pilot Maker, 135
  simulator companies, 72
  Sports Illustrated profile, 95
  submersibles, 107–9, 112, *134*
  Trainer 50th anniversary, 120
  Trainer demo for Army Air Corps, 37
  Watson on, 77
  work at piano factory, 26
  See also Celestial Navigation Trainer;
    flight simulation; Ocean Systems
Link, Edwin A., Sr., 33
Link, George, 33–34
Link, Marilyn, 110

Link, Marion, 33, 34, 37, 96–99, 120
Link Aviation
  Abbey and, 77
  commemoration, 129–30, 133, *142–43*
  formed, 33
  Kelly on, 126
  Lash on, 89
  selling off, 72
  See also Link Trainers; space flight
    simulation
Link Piano and Organ Company, 18–20,
    33, 129–30, 133
Link Trainers
  overview, 136–38
  50th anniversary, 120
  Black people as instructors, 46, 47
  Black people training with, 50
  Celestial Navigation Trainer, 50, 51–52,
    54, 67–68
  at circuses, 34
  computers and, 60–61
  Deep Submergence Systems Review
    Group, 105
  demo for Army Air Corps, 37
  as depth technology, 136
  diagrams, *10, 30, 74*
  as entertainment, 31, 33–34, 67–68
  GAT-1, 92–93
  government order for, 6
  government requirements and, 71
  history of, 7
  instruments, 44
  Marilyn and, 110
  player pianos and, 83
  as psychological training, 60–61
  restoring, 88–89, 92, 120–21
  selling off, 72
  sold to Japan, 37
  suspension of disbelief and, 37–38
  trusting instruments, 40–41
  VAMP, 61
  women as instructors, 46
  workings of, 31–33
  See also Apollo simulator program;
    Mercury simulator
Linkanoe, 95
lions, 15
Lovell, Jim, 86–87
low-tech simulation, 115–16
loyalty, 125, 126, 129
lunar lander, 83

MacInnis, Joe, 105–6, 109–10, 113
maintenance, 92
Marconi, Guglielmo, 131
Mazzoni, Carl, 89–90
McKee, Art "Silver Bar," 138
mechanization, 18. *See also* pianos/organs
Mercury simulator, 77
Microsoft Flight Simulator, 63
military simulation, 116–19
Minor, C. Sharpe, 19
Miss Doran, 13, 14
Mitchell, Billy, 36
Mock, Ron, 65–66
modernity, 92
Monitor, 112
Moroney, William, 115
Morse code, 24, 106
Morton, James, 55–58
motion trainers, 27–28, 64–65. *See also* Link Trainers
motivational fidelity, 64
multilevel thinking, 91–92
music, 18. *See also* pianos/organs

NASA, 75, 78, 83, 85–87. *See also* Apollo simulator program
Nautilus, 101
navy, 104–5, 109–10, 115
A New Theory on Columbus's Voyage . . . (Link and Link), 98
No. 1 British Flying Training School, 1–4
Noonan, Fred, 51
normal engineering, 82
Notice to Airmen/Missions (NOTAM), 23
nuclear power plants, 90–91
nuclear submarines, 104, 105, 108

Ocean Systems, 104
Ocker, William Charles, 28, 29, 38–39
oil industry, 111
Oklahoma, 12, 13
Operation CHASE, 111
organs. *See* pianos/organs
Our Future Is in the Air (Picasso), 20
overspecialization, 139

Pabco Pacific Flyer, 12–13, 14
Peerless Piano Player Company, 16
penguin system, 28
Perry, John, 107
Peters, David, 69

Peterson, Jon, 116
Petrick, Jason, 69–70
petteia, 115
physical fidelity, 64
pianos/organs, 16–19, 83, 88, 120. *See also* Link Piano and Organ Company
Picasso, Pablo, 20
Piccard, Auguste Antoine, 101–2
Pigott, Peter, 20
Pilot Maker, 135
pilots. *See* aviation; flight contests; flight simulation; flight training; Link Trainers; low-tech simulation
Port Royal, Jamaica, 98–99
Powder Puff Derby, 45
Powell, William Jenifer, 48–49
Project Gemini, 77, 80
Project Mercury, 77, 80
pucker factor, 65

Queen Bess, 48

radical design, 82–83
radio, 131
Reagan, Ronald, 127
Richards, Jonathan, 70
Rickenbacker, Eddie, 36
Rickover, Hyman, 104–5
Roosevelt, Eleanor, 49–50
Roosevelt, Franklin Delano, 2, 36–37, 42, 46–47
Roosevelt, Theodore, 115
Royal Air Force, 1–4
The Royal Air Force in Texas (Killebrew), 2–3
Ruggles, William Guy, 28

Salas, Eduardo, 72
Sanders, Haydn, 26–27
Schaff Brothers Piano Com-, 18
Schluter, Paul, 14, 15
Schneider Trophy, 39
Schneider, Timothy, 38
Scrape project, 83
Sea Diver, 97, 99, 106
Sea Diver II, 99–100
Seamans, Robert, 75
segregation, 3, 47
Seiden, Alex, 68
sextants, 50
shatranj, 115

Shepard, Alan, 77
Sherwood, Susan, 87–89, 92
shipwrecks, 96, 113
Siegel, Greg, 60
sight, 38–39
SIMNET, 117–19
simulation, 69–71, 72, 77, 90–92, 115, 136.
    See also Apollo simulator program;
    flight simulation; low-tech simulation
simulator networking, 117–19
Singer Corporation, 72
Sky Chief service, 41–42
sky roadways, 21–22
Snook, Mary, 45
Soviet Union, 77
space flight simulation, 77, 80–81. *See also*
    Apollo simulator program
space programs, 75–77, 80. *See also* NASA
space travel ideas, 79–80
Spirit of Peoria, 12
Sputnik, 77
steam power, 120
Stearman plane, 3–4
Sténuit, Robert, 103, 106
stimulators, 115
Stultz, Wilmer, 45
submarines, 101, 104–5. *See also* nuclear
    submarines
Submersible Portable Inflatable Dwellings
    (SPIDs), 103–4, 106–9
surveillance, 118
Swigert, John, 86–87
SWIP project, 83
systems engineering/thinking, 84–85, 90,
    91–92
systems management, 84

TechWorks! 88, 92
Telford, Thomas, 121
ten C's, 52
Terrell, Texas, 1–4
Thaden, Louise, 45
THINK, 124–25
Thorpe, Jack, 116–19
Three Mile Island, 90–91
Thresher, 104, 105
Titanic, 113
toxic waste, 111
Transcontinental and Western Air (TWA)
    Flight 6, 41–42
transfer of training, 63

Trieste, 102, 104
Turtle, 101
Tuskegee Airmen, 49–50

underwater capsule, *94*
underwater living, 101–7
United Airlines, 65

Variable Anamorphic Motion Picture
    (VAMP), 61
Vaucanson, Jacques de, 17
Verne, Jules, 79–80
vertigo, 64–65
virtual environment, 32. *See also* flight
    simulation; low-tech simulation;
    simulation
virtual reality, 32, 136
vision, 38–39
visual effects, 68, 71
Visualizing Software (Bennett), 91
Vonnegut, Kurt, 88, 92

Ward, Earlin, 90
wargames, 115–19
Watson, Maurice, 55–58
Watson, Tom, Jr., 77, 126
Watson, Tomas, Sr., 124–26
weather
    airmail and, 35–36
    aviation, 6, 14–15
    Guggenheim fund, 39–41
    Link Trainers, 52
    TWA Flight 6, 41–42
Webb, James, 75–76
Weems, Philip Van Horn, 50–51, 97
Whitney, Joshua, 15
Wiesner, Jerome, 75
Wilt, Donna, 62–64, 66–68, 72
women, 5, 44–47, 48
Women Accepted for Voluntary Emer-
    gency Services (WAVES), 46–47
Women Airforce Service Pilots (WASP),
    47
Woolaroc, 13–14, 51
working academy, 121
World War I, 47
World War II, 1–2, 46–47, 49–50
Wright, Orville and Wilbur, 20, 26–27, 31,
    136, 137

Young, John, 80

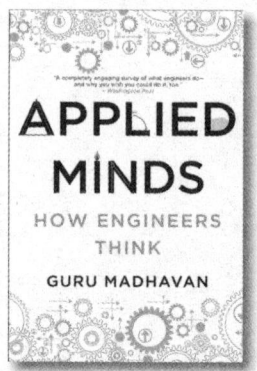